✷ Cleanse ✷ Nourish ✷ Rejuvenate ✷ Heal

Wheatgrass

Nature's Finest Medicine

The Complete Guide to Using Grass Foods & Juices to Help Your Health

by Steve Meyerowitz

Sproutman®

✶ Nutrition ✶ Benefits ✶ Research ✶ History & Religion ✶
✶ Growing ✶ Juicing ✶ Healing ✶ Detoxification ✶ Retreats ✶
✶ Testimonials ✶ Resources ✶ Chlorophyll ✶ Barley Grass ✶

Printed in the United States
of America.

ISBN 1-878736-72-8 Paperback

Newly Revised and Expanded
FIFTH EDITION

Special thanks to the Int'l Biogenic Society for permission to extract from the Essene Gospel of Peace by E.B. Szekely. And to the Ann Wigmore Foundation for permission to use her words in the Epilogue.

Sproutman Publications
PO Box 1100, Great Barrington, Mass. 01230
413-528-5200. Fax 413-528-5201. Http://Sproutman.com

Table of Contents

Dedicated to Charles F. Schnabel

Dr. Charles F. Schnabel
The Father of Wheat Grass
1895–1974

In 1930, Charles F. Schnabel started feeding his family grass. Before anyone else, he initiated the movement for the human consumption of grass and devoted his entire life to promoting its nutritional and health benefits. He also succeeded in creating a demand for grass as a premium livestock feed and furthered its role as a profitable and ecological crop for American farmers. His dream was to see grass included as a valuable supplement in the American diet. He knew from his experiments with animals and his research in the laboratory that it boosts nutrition, builds good blood and strengthens immunity. His vision was of an America that would donate grass to feed the hungry worldwide and teach malnourished countries to grow it. He is a forgotten hero, but he is remembered in these pages for how close he came to making wheat grass a household food. Few people are aware of it today, but in the 1940's, pharmacies all over America and Canada sold 'tins' of grass. Stories about the human consumption of grass appeared in Newsweek, Business Week, Time and other magazines. Today, grass is just now approaching the level of popularity that Charles Schnabel had crusaded for and achieved over sixty years ago.

This book is dedicated to Dr. Charles F. Schnabel, chemist and agriculturalist, for his lifelong and tireless efforts to promote the nutrition and health benefits of grass. May his dream come true.

A Wheatgrass Primer

What's in this Book?

What is Wheatgrass?

Wheatgrass is a variety of grass that is used like an herbal medicine for its therapeutic and nutritional properties. It is available as a fresh squeezed juice, a dehydrated powder, or tablets. This book uses the name "wheat" grass because it is the most popular, but the common grains of barley, oat and rye grow grasses that are equally potent. *See Spiritual & Religious Roots, History & Culture, Healing with Grass, Nutrition, The Pioneers, Real Stories from Real People.*

What does it do?

It has broad effectiveness, but its three most therapeutic roles are: blood purification, liver detoxification, and colon cleansing. As a food it is very nourishing and restorative with such a complete range nutrients that it can, by itself, sustain life. This nutritional miracle is most evident in the animal kingdom where studies prove large and small grazing animals not only sustain themselves on young grasses but also improve their health. *See Healing with Grass, Nutrition, Research, Real Stories from Real People, Spiritual & Religious Roots.*

How do I take it?

Therapeutically, you would drink the fresh juice or apply it rectally through enemas or implants. For nutrition and prevention, you can make powdered drinks or take tablets. *See The Juicers, Healing with Grass, Real Stories from Real People.*

Where do I get it?

From your natural food store, juice bar, direct from growers, or mail order. *See Resources, The Companies, The Pioneers, Healing Resorts.*

Why should I take it?

Wheatgrass earned its reputation from people with terminal illnesses who took it at the eleventh hour of their lives, after conventional medicine left them with no hope. But you can take it as part of a long range prevention and health maintenance program. *See Healing with Grass, Research, Nutrition, Real Stories from Real People.*

How do I get started?

You can grow the grass yourself, buy it from a grower or health food store, drink the juice at a juice bar or buy bottled grass tablets and powders. But if you are sick, it is highly recommended that you enroll in a retreat center for a 2–4 week wellness program. As an alternative, you can establish a home-health program using the information in this book and the guidance of a knowledgeable health professional. *See Grow Your Own Grass, The Juicers, The Companies, Healing Resorts.*

Why should I believe you?

There are many scientific studies demonstrating the efficacy and nutrition of grass foods. Most information about its therapeutic effectiveness is based on clinical evidence and the word-of-mouth testimony of users. *See Science & Wheatgrass, Research, Real Stories from Real People, Nutrition, Spiritual & Religious Roots, History & Culture.*

Wheat Grass vs. Wheatgrass

A word on spelling. "Wheat grass" is a variety of grass like barley, oats and rye, grown in fields across America. "Wheatgrass" refers to grass grown indoors in trays for approximately 10 days and is the kind that is squeezed into a fresh juice. The tray-grown grass is used primarily for therapeutic purposes. The 60+day old field grown grasses, available in dehydrated powder or tablets, are used primarily as nutritional supplements.

Disclaimer

The information in this book is not intended to be a prescription for the user. This book does not claim to cure anything; it does not offer medical advice. Instead it presents research, personal experiences, testimony and nutritional information regarding the use of plants and vegetables in harmony with natural laws for healing and rejuvenation. Please do not use this book if you are unwilling to assume the sole responsibility of evaluating and choosing your own diet/lifestyle or treatment program. Because each person's health condition is unique, the author urges the reader to consult a qualified health professional before undertaking any suggestions described in this book. Good luck and good health.

Acknowledgments

Although my name is placed on the cover of this book, the information presented within comes from a long list of contributors: pioneers, entrepreneurs, researchers, manufacturers, and a few saints. I am grateful to every grower and purveyor mentioned in these pages for their support of my efforts to provide good information to you. What follows is a few special thank you's and the inevitable risk of leaving someone special out.

Thank you to Ron Seibold and the staff of *Pines International*, "the wheat grass people." More than anyone, they carry the torch handed down by Dr. Charles Schnabel of promoting the value of cereal grasses. They do it with quality and integrity. Pines helps return thousands of acres of denatured land to organic soil and donates millions of dollars worth of grass foods to feed the hungry in third world countries. Schnabel would be proud.

Thank you to the *Green Foods Corporation* and the gracious and informative Dr. Bob Terry. Green Foods Corporation has provided a wealth of nutritional and scientific research on barley grass. Founded by Dr. Yoshihide Hagiwara, a living grass foods legend, this company has a strong medical and pharmacological foundation. Hagiwara puts his profits back into research and has added a credibility to grass foods that only science can offer. The entire industry benefits from his work which, in its absence, would depend largely on testimonials.

Thank you to Emily Schnabel-Sloan who took the time and energy to dig through boxes of old papers of her father's work. Thank you to Dick Houston who worked with Dr. Schnabel. Thank to Julie Irons and her gloriously long memory and to son Robert Irons. Thank you to the *Optimum Health Institute* and the generous and enthusiastic Robert Ross whose photos of wheatgrass grace many of these pages. A thank you to Piter Caizer the "Wheatgrass Messiah," who spreads the message about wheatgrass through his music wherever he goes, and who shines with the powerful energies that grasses and living foods provide. A thank you to the individuals whose experiences with wheatgrass in their personal healing are included here for all to see.

Thank you to all the librarians at the Simons Rock College Library in Great Barrington, Massachusetts and to my assiduous proofing editors Nancy Flaxman, Pauline Clarke and my wife Beth Robbins whose support system was so valuable to me in laboring on this book.

While I cannot thank all the deserving professionals here, I want to take a moment to comment in general on the businesses in the grass foods and natural products industry. Small manufacturers rarely get their time in the spotlight. Operating in the manic and sometimes ruthless marketplace of the 1990's, consumers can easily develop defensive and even cynical attitudes towards anything commercial. The mere word "corporate," for some, is enough to generate angst. Corporate America has earned this reputation because it raped our lands with dioxin and DDT, polluted our rivers with PCB's, sprayed our vegetables with pesticides and blackened our lungs with cigarettes. Profits before people is its indelible image.

But in direct reaction to this poor record, many younger companies have emerged with purer principles and missions for creating a better world. Mostly, these companies are formed by visionaries. Integrity matters to them. These are real people with a real desire to make a contribution. They have a dream and their labor is not just about money. I have met many of these entrepreneurs whose dedication to personal and planetary health is exciting. Business is a motivating force in our society—good or bad. It can and has convinced us of the desirability to smoke Marlboro's or eat yoghurt. It is particularly gratifying to see companies promoting back-to-nature concepts and organic whole foods. Not all companies make lots of money—far from it. Small companies spend most of it on research, equipment, processing, packaging, advertising, marketing and employees. When all is said and done, most are not rich. Even the giants in the natural food industry are little fish compared to the whales of mainstream commerce. These hard working natural product companies are helping us improve the quality and the quantity of our lives. To them, I say "thank you."

Steve Meyerowitz, Great Barrington, Massachusetts. 1998.

History and Culture

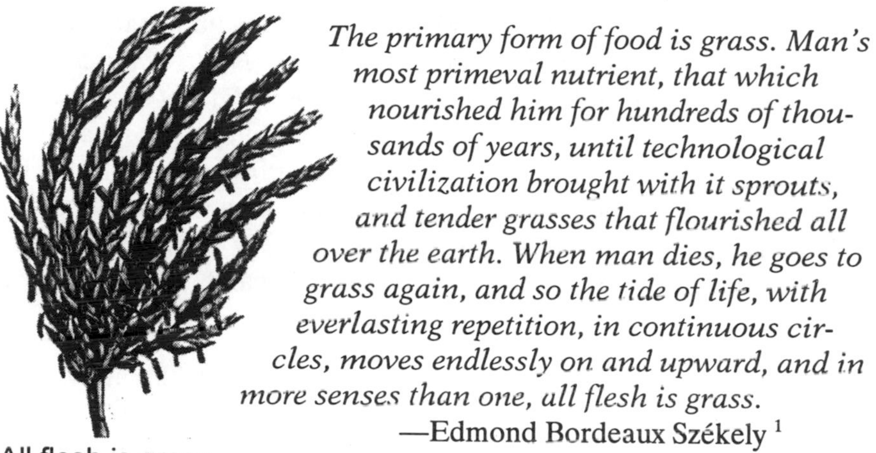

The primary form of food is grass. Man's most primeval nutrient, that which nourished him for hundreds of thousands of years, until technological civilization brought with it sprouts, and tender grasses that flourished all over the earth. When man dies, he goes to grass again, and so the tide of life, with everlasting repetition, in continuous circles, moves endlessly on and upward, and in more senses than one, all flesh is grass.

—Edmond Bordeaux Székely [1]

All flesh is grass

We step on it, sit on it, lay on it, jog on it, picnic on it, walk the dog on it, mow it, water it, in fact, we do most everything on it, for it, or with it except eat it! Wherever there is sun, water and earth, there is grass. From the outback down under to the one inch Arctic tundra of Greenland (they call it Greenland) to the hundred foot tropical bamboo, grass is the most fundamental form of vegetation on the planet. Only algae and lichen grow in the extremes of climate that grasses can survive. There are over 9,000 known species of the grass or grain, "gramineae" family. Wheat alone is cultivated on one third of the planet's farm land and grains in general account for half of the world's agriculture.

Grass arrived long before humanity and will undoubtably remain after we have gone. The highest civilizations of the past have coincided with the best grass lands. The Egyptian goddess of fertility Isis, is purported to have discovered the wheat grain in Phoenicia (now Lebanon). The famous Greek historian Herodotus, described this area, the eastern shore of the Mediterranean sea, as the fertile crescent. It is the cradle of Western civilization and a land of unbelievable fertility. Many of our cereal grasses originated here. Ceres (cereal) is what the Romans named the goddess of agriculture. The Greeks used grains as gifts to their goddess of harvest, Demeter and her daughter Persephone. The Chinese honored the cereal grains with elaborate ceremonies conducted as early as 2,800 B.C. The ancient shepherds were nomads and followed the grass season. The prophet and shepherd Isaiah knew how important grass was to humanity: *"All flesh is grass and its beauty is like the flowers in the field."*[40/6] And the prophet Jeremiah: *"Their eyes did fail, because there was no grass."*[14/6] Jesus told his followers the humble grass held secrets of heaven and earth and gave birth to all creation. *(See p. 19)* Nebuchadnezzar, King of Babylon, was pitied as a madman when he forsook his palace and ate grass in the meadow. Was he crazy after all? The basic unwritten theme of early human history was the search for greener grass.

What Is Grass?

To a botanist, it is a plant in the gramineae family with narrow leaves, hollow stems and inconspicuous flowers. The leaves are attached to the stems at joints or bulges. Grasses are the most widely distributed group of flowering plants. You'll find them above the Arctic Circle, in temperate and tropical prairies, forests, savannahs, all the way to Antarctica. The "fruit"of grass is grain. Every time you plant a grain, you will grow grass. Wheat, rye, corn, rice, oats, barley, sorghum, millet, spelt, Kamut all make grass. Even the two inch thick sugar cane plant in the tropics and the hundred foot tall bamboo in the Far East are part of the same grass family as the grass on your front lawn. And there are numerous sub-species. Wheat, for example, is part of the genus "triticum," along with oats, barley and rye. It has varieties such as hard wheat, soft, winter, spring, red, gold, durum, semolina, and so forth. Wheat and its sister grains throw out shoots and roots so tightly that they are matted like a green carpet and thus ideal for lawns. When growing it at home in a tray, you can lift the grass out just like a rug. In fact, these root systems can represent up to 90% of the plant's weight[2] enabling grasses to endure drought, cold and even fire. Compare their resilience after a frost to the vegetables in your garden. Can this fortitude be an indicator of its nutritional powers?

Its ubiquity is commensurate with its eminence.

The Cultural Significance of Grass

Grasses are of vast ecological and economic importance. They have been a dominant source of human food throughout history and worldwide. Grains are a concentrated source of carbohydrates, B vitamins, fatty acids, minerals, fiber and protein. They have migrated with our species from ancient times and places to the modern world. In fact, one quarter of the grass species in the northeastern United States today arrived with European settlers.[3] All the world's cereal crops are grasses and four of the world's top five crops are cereals, wheat, corn, rice and barley. The family also provides most of the world's sugar, from sugarcane. Grasses are the primary source of food for domestic and wild grazing animals, which feed on pastures and grasslands. Of the fifteen major crops that feed animals the earth over, ten are grasses. One third of the planet is covered by grass and even in the cemented cities across America, grass fights back through the sidewalk cracks. Its ubiquity is commensurate with its eminence. Nevertheless, we largely ignore it. We chase after the azaleas, roses and orchids instead.

Modern History

Prior to the late 1800's, grass was just known for being a good livestock feed. But farmers have long observed a qualitative difference in the pelts of livestock that pastured on the young grasses of the early spring. Botanists studied grasses to determine which varieties would produce the healthiest cattle with the highest quality milk, butter and cheese. Animals don't know anything about vitamins. They determine the nutritive value with their instinct, palate and olfactory faculties acting for them in place of judgement. *"The preferences shown by cattle are better proofs than those obtained from the analysis of the chemist,"* said George Sinclair in 1869.[4]

The first documented study of young grass was in 1883 when researchers found that "immature" grass was high in protein and low in fiber.[5] In an analysis that hinted at the jointing theory to be written decades later, an 1890 state agricultural report announced that the mineral matter of grass reached a peak during the period of most rapid growth and declined with maturity.[6] After that, there were no more additions to our knowledge of young grasses on record for forty years. During this time research focused on legumes as the possible answer to our nutritional needs. "Corn is king and alfalfa is queen" was a familiar slogan of the time.

In 1925, an English botanist determined that "young pasture plants are equivalent to protein concentrates." Unfortunately this work lumped together all field grasses, young legumes and young grasses and no distinction was made.[7]

In 1931, Charles F. Schnabel made two discoveries that would change our conceptions about the place of grass in agriculture and initiate the trend for its human consumption. *(See Pioneers, p. 23)* 1: Schnabel demonstrated that a culm of grass reached its peak nutritional value on the day the first joint begins to form. This marks the end of the vegetative stage and the beginning of the reproductive stage of the plant. 2: The food value of the grass at the jointing stage roughly paralleled its protein content. He found that grass grown on richly fertilized land would produce 40% protein grass. This represented a miraculous food value in terms of the agricultural resources and economics necessary to produce protein through other animal and vegetable sources.

In the 1940's, Schnabel inspired the large scale production of young cereal grasses that were dehydrated, canned and sold as nutritional supplements in pharmacies throughout North America. In fact, cereal grass tablets were the nations best selling multiple vitamin and mineral supplements. It wasn't until the early 1950's that it was dethroned by the popular *One-A-day Multiple* vitamins and *Geritol*. These products capitalized on the movement towards technology. "Better living thru Chemistry" was a common slogan of the time, and "man-made" superceded "nature–made" in importance. Science can make bigger fruit with plant hormones, more profits per acre thanks to fertilizers and rid the land of pests with pesticides. These techniques did indeed increase food production, but at the expense of our health and the purity of our soil, water and air.

In the 1970's Dr Ann Wigmore opened Hippocrates Health Institute in Boston, nourishing terminally ill patients back to health with fresh squeezed wheatgrass juice. Today, virtually every natural food store in North America carries a grass foods product whether it be fresh or dried, wheat or barley or Kamut.

> *Of the numerous tribes of vegetables which clothe and embellish the earth, none is more interesting nor more extensively useful than the natural order gramineae or family of grasses... the food of man, as well as that of the more useful animals, entirely depends on the produce of our corn fields and pastures.*
>
> —George Sinclair, 1869 [8]

Spiritual & Religious Grass Roots

Wheatgrass in the Dead Sea Scrolls, Ancient Hebrew & Chinese, Modern Messiah's Spiritual Message

I believe a leaf of grass is no less than the journey-work of the stars.—Walt Whitman[1]

When I touch grass, I touch infinity. It existed long before there were human beings on the earth, and it will continue to exist for millions of years to come. Through the grass, I talk to the Infinite, which is only a silent force. This is not a physical contact. It is not in the earthquake, wind, or fire. It is the invisible world of nature of which I, too, am a part.

—George Washington Carver[2]

Grass, we have seen, has had a long and essential relationship with humans and animals among various cultures throughout history and across the globe. But as we explore deeper into ancient times, cultural and historical records gradually merge with religious history and even mysticism. This chapter establishes a foundation for the relevance and importance of grass in modern times by exploring its religious, mystical and spiritual roots. Specifically, we will look at the definition of grass in the characters of the ancient Hebrew and Chinese languages, listen to Jesus discuss grass with his disciples and hear a modern day "messiah's" view on the spiritual powers of grass.

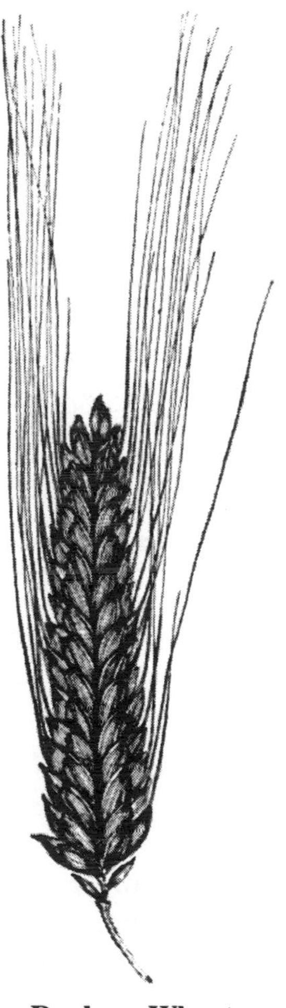

Durham Wheat

The Meaning of "Grass" in Ancient Hebrew and Chinese

Language is very telling. Etymologists spend entire careers tracing a language's original meaning and form. At first blush, it is tempting to assume their work belongs to the domain of museums and Universities, but there are often incredible treasures locked inside language that reveal ancient wisdom and enlighten us, even in today's age of progress and technology.

Ancient Chinese was derived from wall pictures, or hieroglyphs, either painted or carved. Some epigraphs on shells go back 3,500 years. But the first cataloging of Chinese symbols into a language began about 2,800 years ago. A single distinctive symbol represents each word in the language and it takes a vocabulary of about 10,000 characters to read a newspaper. The word for grass in Chinese is a picture of two people kneeling at an alter and praying with heads and arms raised skyward.[3]

Ancient Hebrew, the language in which most of the old testament of the Bible was written, has records from the epoch of King David and Solomon, some 3,200 years ago. Prior to that, the language had roots in Phoenician and Canaanite scripts and hieroglyphs.

In Joe Sampson's book, *Written By the Finger of God—Decoding Ancient Languages*[4], he describes Hebrew as the word of God handed down with a sacred message in each and every character. The Hebrew characters, he says, contain an incredible amount of compressed and layered meaning—a literal spiritual code distilled in every character. The oldest form of the Hebrew word grass is *"Deshe"* spelled דשא, ד *Dalet,* ש *Shin,* א *Aleph.* One interpretation of this Hebrew code would be:

Through the lowest of the kingdoms on earth is
the doorway through which we may be regenerated
and come towards perfection of being.[5]

What is the message encoded in the Hebrew word for grass thousands of years ago? Is it regeneration? A path toward perfection? Is grass the primary herb (Esev) for healing humanity? Modern researchers have defined it as a complete food. Can people live on grass? Why does the green life–giving nourishment trapped inside its blades provide health and longevity to guinea pigs, rabbits and hens while a diet of alfalfa, lettuces and other greens leave them emaciated and sick?[6] How can 2,000 pound grazing animals, cattle, deer, horses, bison, buffalo, antelopes, elephants, gazelles and giraffes all live and thrive on an exclusive diet of grass? Will grass provide the nourishment people need as well? Can what the chemists call the "unidentified factors" or "grass-juice factors" promote our recovery from sickness and disease? Can the blood rejuvenating chlorophyll held within its leaves restore our natural health and longevity?

The Spiritual Powers of Grass

A Conversation with Piter U. Caizer, The Wheatgrass Messiah

Piter U. Caizer is a German born musician who was disheartened by the excessive use of drugs and alcohol in the rock music world. These influential celebrities who reach millions with the universal language of music, were too important to present the wrong example to others or destroy themselves. So Piter decided to do something about it. He brought his juicer wherever he would play and made green juices for everyone. "After playing for hours, musicians get dizzy and tired. So instead of doing drugs and going down, they're doing live foods and going up." He turned everyone "on" and they experienced the long-term energy boost of live green juice. The musicians called him "the Wheatgrass Messiah." His special perspective of why wheatgrass works and its spiritual effects are spoken here in his own words.[7]

Philosophy of Living Foods

If you think about assimilation of sunlight, humans don't take the energy directly from the sun; we're going through other sources—green plants. We should develop a diet that gets closer to the foods that provide that energy, but instead we have gone in the opposite direction with highly cooked, fried, microwaved, processed food and it's all dead. It doesn't provide any vitality. For me it's about vitality. Food has to give me vitality. When I eat food I want to feel energetic and happy and good. When I eat cooked food, I feel tired. It just doesn't work. You eat dead food and it brings your vibration down and your body has to wrestle with it, to get it out of the system, then you come back up. Then you eat again and it goes down again. So if you eat live food, your vibration doesn't go down; it gets higher and higher and you grow in a spiritual way.

Wheatgrass contains raw chlorophyll. Chlorophyll is condensed sunlight. Since we are light beings, spirit and soul inside solid bodies, the light force vibrates through the physical body. That's the energy you feel. That's why wheatgrass is a spiritual food. It nourishes you on the spiritual level as well as the physical.

Rocket Fuel

When I learned about the grasses and what they can do for you, I said "this is it." Now we have got to find a way to make it palatable because if you give people tray grown wheatgrass, they say...uhh, I don't like this. So I worked on the taste. The first thing I mix in my wheatgrass juice is sunflower sprouts. Because sunflower, if you look at the plant, is an amazing plant. It is almost standing human-like with its radiating big face. Sunflower seed is a complete protein. The wheatgrass doesn't have a lot of protein, so by mixing the two together I make it more complete and wow, what an improvement in the taste! So much better, you get a little nutty taste in there from the sunflower sprouts. Then I add some lettuce, and parsley, ginger, beets, celery and alfalfa sprouts. Put in whatever you want. All of a sudden you have this mixture which tastes great and has the benefits of all the individual ingredients mixed together in this dynamic energy juice—I call it rocket fuel.

If you think about the human body like a car, then I want to give it the best fuel I can get. If you go to the gas station you can choose between premium, regular and low octane. Jet fuel is the highest octane. Well, wheatgrass is like jet fuel for humans. It's the best fuel you can get. And, because it goes right into your system, you can feel it. I give green juice to a depressed person and their depression goes away. How is that possible? Because they get back into that harmonious field that we all have deep inside, and you feel in balance; you feel in touch with Mother nature again. But if you eat all these fried foods, micro-waved, you're going the opposite way. You're going out of touch with our nature and we cannot survive apart from nature. We are a part of Mother nature.

Spiritual Growth

We are vibrating beings with invisible spirit–soul bodies inside us. So when you drink grass juice, what happens? It elevates your vibration and you become more aware that there is something higher going on that is not perceptible with the physical senses but which can be perceived with the higher senses which are dormant within us. We have to make an effort to wake up through meditation and concentration, to activate these spiritual senses which are deep within us. But the way our society runs, everything is all 'outside, outside,' and people spend their time running around wasting their lifetimes making money instead of developing consciousness. When we die the only thing we take with us is our consciousness from when we were in the flesh. The more you can develop that, the more you can actively work on your evolution.

Anti-Aging

We talked about chlorophyll. Chlorophyll is as close to the molecular structure of human blood as anything on the planet. So you can actively revitalize your blood. From an external source comes a raw material that the body can instantly convert into fresh blood. We're only as healthy as our blood is. So, the more you can rejuvenate the blood, the more you slow down the aging clock.

> *First they cook it, then they freeze it, then they microwave it. What is left in it? That's dead. I wouldn't even give it to my dog.*

The older you grow, the more you burn up your natural enzymes and the harder it is to digest your food. That creates all kinds of complications in the body. People have all this accumulated, incompletely digested stuff in there because they can't secrete enough enzymes anymore. The linings of the intestines are all plugged up. If you look at the average diet of what people eat, it's loaded with heavy foods like cheeses and meats...sure you can get away with it for some years but look at the big picture, look at all the people hidden away in the hospitals. And what do they do with them? Instead of relieving their toxicity and cleaning them out, they feed them drugs, making them even more toxic. That whole approach just doesn't work.

Then you have people in their fifties that look young. They have this beautiful energy radiating from them because they are doing that transformation within them. Then you look at others and they look hard and stiff. The difference between youth and age is in the flexibility. Age brings solidity. So when you do the juices and the wheatgrass, you don't experience that because you are constantly replenishing your system with the energy of youth from the young grass. So your system is not growing old as much because it is not burning out as much. It's like a bank account. If you constantly draw money out of it, you are eventually going to go bankrupt. That is exactly what the others are doing to their bodies, they are bankrupting their health. They're eating corrosive food, then their bodies become corroded, then their minds become corroded. They're getting old because they are rusting!

We've got to empty the hospitals. In the Chinese way, the doctor is paid to keep you healthy. Once you get sick, he doesn't get paid because he didn't do his job right. That is what is wrong with our medical profession. They don't teach the people to help themselves and heal. In the whole medical profession there is one word they don't use; it's *heal.*

We are all one. We are all from the same ocean. We are all drops of that same ocean. And we only can come back together in the spirit, in the consciousness; we cannot come back together in our opinions because everyone has opinions but we can come together in the truth. And the truth will always be the truth, it cannot be altered by man. And that is what is so beautiful about the truth. No one can change the truth. The truth will always be the truth. And what is the truth about sprouts and wheatgrass will always be the truth, now and in a thousand years.[8]

> *The superior physician helps before the early budding of disease...To administer medicines to diseases which have already developed is comparable to the behavior of those persons who begin to dig a well after they have become thirsty.* –Huang Ti, 2697-2597 B.C. legendary Chinese ruler and cultural hero known as"The Yellow Emperor" who co-founded Taoism with Lao-Tsu.

Jesus' Words from the Essene Gospels

The most precious gift of your Earthly Mother is the grass beneath your feet...
—Jesus, the Essene Gospel of Peace IV [9]

The Essene Gospels represent an exciting piece of history and a timeless message of healing, peace and love. The Essenes were a Jewish sect who lived with Jesus near the Dead Sea. They preserved his words on leather scrolls which were left behind in caves and found generations later after they had fled from Palestine. Some of these scrolls were excavated between 1946 and 1951 and are called the *Dead Sea Scrolls.* The Gospels are translated directly from the Aramaic dialect spoken by Jesus. Thus the translation has a purity and simplicity that remains powerful even in modern times. The following, extracted from the *Discovery of the Essene Gospel of Peace,* by Dr. Edmond Bordeaux Székely, describes in part, the story of how the scrolls were left in the Dead Sea:

"They sent out healers. And one of them was Jesus, the Essene. He walked among the sick and the troubled, and he brought them the knowledge they needed to cure themselves. Some who followed him wrote down what passed between him and those who suffered and were heavy-laden. The Elders of the Brotherhood made poetry of the words, and made unforgettable the story of the Healer of Men, the Good Shepherd. And when the time came at last for the Brothers to leave the desert and go to another place, the scrolls stayed behind as buried sentinels, as forgotten guardians of eternal and living truth.

"A dark age began, a time of savagery, of barbarism, of book-burning, of superstition and worship of empty idols. The gentle Jesus was lost forever in the image of a crucified God; the Essene brothers hid their teachings in the minds of the few who could preserve them for their descendants, and the Scrolls of Healing lay neglected beneath the shifting shadows of the desert..."

About Edmond Bordeaux Székely

Edmond Bordeaux Székely is a well known translator and philologist (scholar of languages). He was a professor of Sanskrit, Aramaic, Greek and Latin, and spoke ten modern languages. His grandfather, Alexandre Székely, was the eminent poet and Unitarian Bishop of Cluj. He is also a descendant of Csoma de Körös, the Transylvanian who compiled the first English-Tibetan dictionary. In addition to translating selected texts from the Dead Sea Scrolls, Edmond B. Székely spent half of the 1920's secluded in the Secret Archives of the Vatican. There, as the result of limitless patience, faultless scholarship and unerring intuition, he discovered and translated the Aramaic scrolls known today as the Essene Gospels.

The first book of the *Essene Gospel of Peace*, published in 1934, has sold over 5 million copies with no commercial advertisement, and has been translated into 26 languages. Book Two, *The Unknown Books of the Essenes* and Book Three *The Lost Scrolls of the Essene Brotherhood* were published almost 50 years later. Book Four, *The Teachings of the Elect*, from which the following is extracted, was published posthumously in 1981 according to Dr. Székely's wishes. They represent yet another fragment of the complete manuscript which exists in Aramaic in the Secret Archives of the Vatican and in Old Slavonic in the Royal Library of the Hapsburgs in Austria. The poetic style of the translations brings to vivid reality the exquisitely beautiful words of Jesus and the Elders of the Essene Brotherhood. The chapter quoted from here is called *"The Gift of the Humble Grass."* Some of the other chapters are: *The Essene Communions, The Sevenfold Peace, The Holy Streams of Life, Light, and Sound.* A catalog of the complete works of Dr. Székely can be ordered from *The Awareness Institute.*[10] *The International Biogenic Society,*[11] which Dr. Székely founded in 1928 with Nobel Prize-winning author, Romain Rolland, survives today.

Székely was so taken with the reverence Jesus and others devoted to grass, he encouraged its use throughout his teachings. During his life, his workshops were always adorned by the presence of grass pots to "charge" the environment for learning. He called them "biogenic batteries." Students would hold them to absorb their healing, restorative, and life-giving properties. The following extracted text reveals some of the reasons why.

From the Essene Gospel of Peace Book IV

by Edmond Bordeaux Székely

Speaking about the angel of air, water and sun to his Brothers of the Elect round about him, Jesus spoke:

"But of all these, and more, that most precious gift of your Earthly Mother is the grass beneath your feet, even that grass which you tread upon without thought. Humble and meek is the angel of Earth, for she has no wings to fly, nor golden rays of light to pierce the mist. But great is her strength and vast is her domain, for she covers the earth with her power, and without her the Sons of Men would be no more, for no man can live without the grass, the trees and the plants of the Earthy Mother. And these are the gifts of the angel of Earth to the Sons of Men.

"Here is the secret, Sons of Light; here in the humble grass. Here is the meeting place of the Earthly Mother and the Heavenly Father; here is the Stream of Life which gave birth to all creation.

"Behold, Sons of Light, the lowly grass. See wherein are contained all the angels of the Earthly Mother and the Heavenly Father. ....For in the grass are all the angels. Here is the angel of Sun, here in the brightness of the green color of the blades of wheat. For no one can look upon the sun when it is high in the heavens, for the eyes of the Son of Man are blinded by its radiant light. And it is for this that the angel of Sun turns to green all that to which she gives life, that the Sun of Man may look upon the many and various shades of green and find strength and comfort therein. I tell you truly, all that is green and with life has the power of the Angel of Sun within it, even these tender blades of young wheat.

"And so does the angel of Water bless the grass, for I tell you truly, there is more of the angel of Water within the grass than any of the other angels of the Earthly Mother....

"Know also, that the angel of Air is within the grass, for all that is living and green is the home of the angel of Air. Put your face close to the grass, breathe deeply, and let the angel of Air enter deep within your body. For she abides in the grass, as the oak abides in the acorn, and as the fish abides in the sea.

"It is the angel of Life that flows through the blades of grass into the body of the Son of light, shaking him with her power. For the grass is Life and the Son of Light is Life, and Life flows between the Son of Light and the blades of grass, making a bridge to the Holy Stream of Light which gave birth to all creation...

"The Earthly Mother is she who provides for our bodies, for we are born of her, and have our life in her. So does she provide for us food in the very blades of grass we touch with our hands. For I tell you truly, it is not only as bread that wheat may nourish us. We may eat also of the tender blades of grass. That the strength of the Earthy Mother may enter into us. But chew well the blades, for the Son of Man has teeth unlike those of the beasts, and only when we chew well the blades of grass can the angel of Water enter our blood and give us strength. Eat, then, Sons of Light, of this most perfect herb from the table of our Earthly Mother, that your days may be long upon the earth, for such finds favor in the eyes of God.

"I tell you truly, the angel of Power enters into you when you touch the Stream of Life through the blades of grass. For the angel of Power is as a shining light that surrounds every living thing just as the full moon is encircled by rings of radiance and as the mist rises up from the fields when the sun climbs in the sky.....

"Touch, then, the blades of grass and feel the angel of Power enter the tips of your fingers, flow upwards through your body, and shake you till you tremble with wonder and awe.

"Know, also, that the angel of Love is present in the blades of grass, for love is in the giving, and great is the love given to the Sons of Light by the tender blades of grass. For I tell you truly, the Stream of Life runs through every living thing, and all that lives, bathes in the Holy Stream of Life. And when the Son of Light touches with love the blades of grass, so do the blades of grass return his love, and lead him to the Stream of Life where he may find life everlasting....

"Touch the blades of grass, Sons of Light, and touch the angel of Eternal life. For if you look with the eyes of the spirit, you will truly see that the grass is eternal. Now it is young and tender, with the brightness of the newborn babe. Soon it will be tall and gracious, as the sapling tree with its first fruits. Then it will yellow with age, and bow its head in patience, as lies the field after the harvest. Finally, it will wither,...but it does not die, for the brown leaves return to the angel of Earth, and she holds the plant in her arms and bids it sleep, and all the angels work within the faded leaves and lo, they are changed and do not die but rise again in another guise. And so do the Sons of Light never see death, but find themselves changed and risen to everlasting Life.

"And so does the angel of Work never sleep, but sends the roots of the wheat deep into the angel of Earth, that the shoots of tender green may overcome death and the reign of Satan. For life is movement and the angel of work is never still....Touch the blades of grass, and thereby touch the Stream of Life. Therein you will find Peace, the Peace built with the power of all the angels. Even so with that Peace will the rays of Holy Light cast out all darkness.

"When the Sons of Light are one with the Stream of Life, then will the power of the blades of grass guide them to the everlasting kingdom of the Heavenly Father. And you shall know more of those mysteries which is not yet time for you to hear. For there are other Holy Streams in the everlasting kingdoms; I tell you truly, the heavenly kingdoms are crossed and crossed again by streams of golden light, arching far beyond the dome of the sky and having no end. And the Sons of Light shall travel by these streams for ever, knowing not death, guided by the eternal love of the Heavenly Father. And I tell you truly, all these mysteries are contained in the humble grass, when you touch it with tenderness and open your heart to the angel of Life within.

"...For so were your fathers taught of old, even our Father Enoch. Go now, and peace be with you."[12]

Grass is the Forgiveness of Nature

by John J. Ingalls, 1898 [13]

Grass is the forgiveness of nature. Lying in the sunshine among the buttercups, daisies and dandelions of May, our earliest recollections are of grass. Grass is the forgiveness of nature, her constant benediction.

....It's tenacious fibers hold the earth in its place and prevent its soluble components from washing into the wasting sea. It invades the solitude of deserts, climbs the inaccessible slopes and the forbidding pinnacles of mountains, modifies climates, and determines the history, character, and destiny of nations. Unobtrusive and patient, it has immortal vigor and aggressiveness.

Banished from the thoroughfares and the fields, it abides its time to return, and when vigilance is relaxed or the dynasty has perished, it silently resumes the throne from which it has been expelled, but which it never abdicated. It bears no blazonry of blooms to charm the senses with fragrance or splendor, but its homely hue is more enchanting than the lily or the rose. It yields no fruit in earth or air and yet, should its harvest fail for a single year, famine would depopulate the world.

Jeremiah 14/6: "Their eyes did fail, because there was no grass."

Isaiah 40/6: "All flesh is grass and its beauty is like the flowers in the field."

The Pioneers

Charles Franklin Schnabel
The Father of Wheat Grass

15 lbs. of wheat grass is equal in overall nutritional value to 350 pounds of ordinary garden vegetables. We have not even scratched the surface of what grass can mean to man in the future.[1] —Charles F. Schnabel

July 31, 1930 was an important day in the history of grass foods. On that day, Charles F. Schnabel got 126 eggs from 106 hens. Anyone who knows about chickens will tell you that those are phenomenal results. To make it even more of a tall story, the hens were sick and dying when Schnabel got them. He only took them to save them from extermination. What does this have to do with grass? To restore their health, Schnabel fed them a mixture of fresh cut, young oat grasses and greens. His miraculous results inspired him to test it on himself. He dried the young greens on his wood stove, powdered them and added them to his family's meals. He had a large family and everyone, including relatives and neighbors, were fed Schnabel's grass. To this day, his daughter still remembers the smell of drying grass that filled the entire house. Such was the beginning of the human consumption of grass in modern times. From this point forward, man has continuously consumed young grasses. Commercial enterprises, thanks to Schnabel, have produced and packaged young grasses. This was the beginning and this was the man who started it.

Schnabel was an agricultural chemist. He specialized in soil fertility, animal feeds and protein research. In the 1920's he worked for the Standard Milling Company and the Kansas Department of Agriculture. So when the depression came and he was laid off in 1930, he had the time and the skills to nurse dying hens back to health. The next spring, he repeated his tests with the hens. Again, the results were striking. Egg production doubled. The eggs had stronger shells and the newborn chicks were free of the common diseases. Even a child could see the difference in their feathers. The grass eating hens averaged 89% productivity while the alfalfa hens averaged 40% productivity. These hens received the same mash and grain ration as the others but with alfalfa added instead of grass.

No one ever heard of such consistently high productivity. Not only that, the mortality rate of the alfalfa hens was ten times greater than the grass hens. In the laboratory he found out why: "The most striking thing was the appearance of their livers which were a dark mahogany color and the surface glistened like a mirror. The alfalfa fed hens had light tan colored livers. These liver changes caused by good grass are too obvious not to have some connection with the prevention of degenerative disease."[2]

Schnabel knew that there was something really potent in the young grass and he immediately started promoting his discovery to feed mills, chemists and the food industry. In the years that followed, he saw his experiments confirmed in the laboratory with chickens, turkeys, rats, rabbits and guinea pigs. Research chemists from major universities identified all known vitamins in grass (except vitamin D) including the newly discovered vitamin K and unnamed nutrients known as the "grass juice factors." *(See chapter Nutrition)* His discovery of the "jointing stage" or the ideal time to harvest grass for maximum nutrition was approved for patents by the USA Patent Office. Large corporations such as Quaker Oats and American Dairies Incorporated invested millions of dollars in fur-

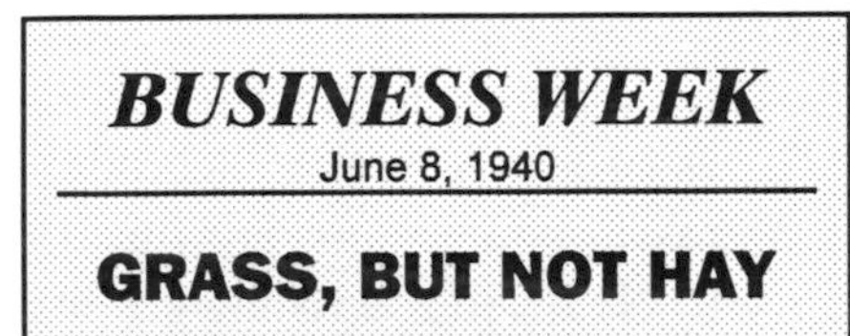

BUSINESS WEEK

June 8, 1940

GRASS, BUT NOT HAY

Butchers, bakers, growers and processors of nearly every description of foodstuff have been finding green pastures under the spreading vitamin tree. And, as everybody must know by now, the green pastures are no longer figurative, since grass itself has shot up to the dignity of a health food.

Cerograss and Cerophyl–the trade names for dehydrated cereal grasses—have received such reams of unsolicited publicity that, novelty or no novelty, business is taking notice.

ther research and the production of products for animals and humans. Rockhurst College bestowed on him an honorary Doctorate of Science for services to humanity. Bakers made grass enriched breads; the AMA approved grass as a "food," and by 1940, just ten years after his sick hens recovered, cans of Dr. Schnabel's grass were for sale in major drug stores all across the United States and Canada.

Ultimately however, the commercial success of his grass would be short lived. Cerophyl Laboratories of Kansas City, Missouri was the company formed by Quaker Oats and American Dairies for the sale of young cereal grass for human consumption as a drug store nutritional supplement and as a livestock and poultry feed. The name was an abbreviation for chlorophyll from cereal grains. The Green Melk Company, Ltd. was their Canadian counterpart headquartered in Guelph, Ontario. More than 100 persons were employed in Kansas City. But the end of WWII created a widespread materialistic euphoria that diluted public emphasis on personal nutrition. Quaker Oats dropped out, and by 1945 Schnabel was seeking new associations. Unfortunately, the financial and marketing clout of large corporations is required to maintain a product in the national arena and Schnabel never replicated that commercial success.

Despite grass' mass market distribution, Schnabel didn't make much money. He was not in it for the money. His goal was to eliminate hunger and malnutrition and he believed grass could do it. His plan proposed a 4 part system: 1. The soil must be reclaimed and enriched. 2. A network of nutritional grass farmers would grow 40% protein grass at greater profit than other crops. 3. A new industry would emerge to properly process, dehydrate, bottle and distribute the grass. 4. Consumers would benefit from a low cost naturally balanced concentrated

Bread of Grass. Still Going Strong

Zinsmaster Baking Co. put the emerald-colored loaf on the market in October and reported that first-week sales were the best ever recorded by the firm for any new product. This bread contains virtually every known vitamin except D with a whole alphabet of those undiscovered in its grass-juice factor. The bread was first offered in Duluth where it went like wildfire. The powder also called Meadow-Kist, is made by Doughboy Mills, Inc. and is packaged as a drug-store item, too. –Business Week 2/1/1941

THE KANSAS CITY STAR

KANSAS CITY, APRIL 8, 1940—MONDAY

GRAZE YOUR WAY TO HEALTH

CHEMISTS AND STAFF THRIVE ON POWDERED GRASS

New Food Has More Vitamins than the Alphabet has Letters

Fortified physically as well as professionally with grass, three Kansas City chemists left today for Cincinnati to tell the American Chemical society's convention that grass is man's best food. Grass is so full of vitamins the alphabet hasn't enough letters to name them all. Four years of experimenting in the Cerophyl laboratories convinced Charles Schnabel, Dick Graham and George Kohler of that.

Meanwhile, for many months, the three chemists have eaten grass and for several weeks the entire laboratory staff of eighteen persons has had grass luncheons. The powdered grass is put in stews, biscuits, muffins, macaroni, noodles and even candy bars. A 3 year test tube attack revealed grass contains almost all vitamins except D which normally is suppled to grass eating animals by sunlight. A parade of vitamins showed up in the grass. Then there was the "grass juice factor" accounting for some food values the vitamins hadn't included. Vitamin K from the grass products was recently found to have blood congealing powers and is being used medically. The paper to be read Wednesday to the chemical convention by Dr. Kohler will claim grass contains vitamins in far richer amounts than fruits and vegetables. A Kansas City obstetrician who used the powders in treating cases of female illness has reported favorably on the effects. At the Mayo clinic, groups of interns have used the powder in experiments. Grass powder as fortification for human diet is definitely proved, Dr. Graham believes and may mean a new cheap way to afford better nutrition.

Chemists say Grass is Man's Best Food (l-r) Dr. WR Graham, CF Schnabel and GO Kohler.

source of vitamins. Schnabel made extensive calculations as to how many farmers it would take, how many acres, the cost per delivery of those nutrients with grass as compared to other foods. He maintained that high protein (40%) grass was the most profitable crop a farmer could grow based on its market value as a fertilizer, an animal feed and a human nutritional supplement.

What is Schnabel's legacy? While he had much success, he also had many struggles. It was hard to preach about such lofty ideas as ending famine and hidden hunger to a post-war, post-depression era public that was bent on consumerism. Schnabel made several presentations before the American Chemical Society, but he was not a writer. There are no books by him and the few papers and pamphlets he wrote were published mostly in small agricultural and nutritional magazines or regional newspapers. He achieved virtually no fame, his name being lost behind the corporations that produced his ideas or the scientists that published the research on them. But the legacy of his life is fourfold.

First, he inspired a body of scientific research on grass that remained unprecedented until the 1980's (*see p.37*). In this, Schnabel was greatly assisted by his colleague, George Kohler. Kohler was a biochemist at the College of Agriculture, University of Wisconsin, Madison. Together with Conrad A. Elvehjem and E.B. Hart, they were the most active group substantiating the food value of cereal grasses in the 1930's and 1940's. This was a prestigious lab. It produced the discovery of niacin (vitamin B3) by Elvehjem in 1936 which prevents the deficiency disease pellagra. Most significant in their research was the discovery of the "grass juice factor" which indicated that the health improvement powers of grass were distinct from any known vitamins. Kohler was so impressed by the potency of young grass that he moved to Kansas City and worked with Schnabel for several years.

Secondly, Schnabel discovered and patented the jointing theory. To just eat grass was not enough. The secrets of grass's nutritional and healing powers have much to do with timing. He pinpointed the exact time that grass achieved its nutritional peak. He proved that if cut a week before or a week after, it had only a fraction of its protein. "My only clue was that the hens refused to eat the older grass."[1] This was later scientifically verified by Kohler.

Thirdly, Schnabel strongly emphasized the need for "organic" farming at a time when the term was not even used. His goal was to grow 40% protein grass. He found 45% protein grass on a well manured farm and produced 30%–40% protein oat grass on his own property. But most farmers' soil was inadequate to produce anything greater than 10%–20% protein. Schnabel pushed for the use of eggshells, seaweed, manure, autumn leaves and recycled grass fibers to enrich the soil. If it's not in the soil, it's not in the vegetable. He understood the value of minerals and trace minerals at a time when agriculture was turning to chemicals as a solution to diminishing crop yields. *"Soil fertility is like money in the bank. To prosper a farmer must put back more than he takes out."*[3]

Although he never repeated the success of the mass market distribution achieved from 1940–1945, his final legacy is the fact that young cereal grasses of oats, barley, wheat and rye continue to be cut at the jointing stage and sold as nutritional supplements, seventy years after his discovery. All of the cereal grass producers in existence today were, either directly or indirectly, spawned by the efforts of Dr. Charles F. Schnabel.

JAMA *Journal of the American Medical Association*

Council on Foods

ACCEPTED FOODS

THE FOLLOWING PRODUCTS HAVE BEEN ACCEPTED BY THE COUNCIL ON FOODS OF THE AMERICAN MEDICAL ASSOCIATION AND WILL BE LISTED IN THE BOOK OF ACCEPTED FOODS TO BE PUBLISHED. —FRANKLIN C. BING, SECRETARY, 1939

CEROPHYL

Manufacturer.—Cerophyl Laboratories (Division of American Dairies, Inc.), Kansas City, MO
Description.—Dried, powdered mixture of young leaves of wheat, oats and barley, selected and blended to maintain the minimum vitamin potency declared on the package label.
Manufacture.—The cereals are grown on soil fertilized to produce plants of high mineral and vitamin content. The young rapidly growing leaves are harvested just before they joint by machinery especially designed to prevent the leaves coming in contact with the ground after cutting. No toxic spray materials are used. The method of cultivation precludes contamination with weeds. The freshly harvested leaves are immediately cut into short lengths and dehydrated. Hot flue gas of minimum oxygen content is drawn from a gas furnace through the drying chamber at an initial temperature of between approximately 800 and 900 C. The high initial temperature, which is quickly reduced the evaporation of moisture from the grass, serves as a flash pasteurization of the surface of the leaves. The entire process of drying requires approximately sixty seconds. The dried material leaves the dryer at a temperature of approximately 120C. The dehydrated leaves are mechanically cleaned, pulverized, stored at –18C. And packed in hermetically sealed cans under nitrogen.

"The new way to feed people is to grow and process 40% protein grass. Grass is the cheapest commercial source of vitamins and minerals in the world for either man or animals. High protein grass production will build up the land faster and at the same time pay larger net profits than any other crop that can be grown. Hunger need not be. A way has been found to banish starvation from the earth. It can be done with a fraction of the land in cultivation. No new frontiers are needed. Every nation now can produce the food it needs." [4 5] —Charles F. Schnabel

V.E. Irons

In 1900 the U.S. was in first place as far as health is concerned out of a total of 93 civilized nations. In 1920, we were in 2nd place. In 1978, we had dropped to 79th place. More than 1 notch per year. Think of it, the richest nation in wealth, knowledge, hospitals, doctors and nurses, yet near the bottom in health—all in 82 years. Yet, in 1920, we had no vitamins. A vitamine means–a live amino acid. Today the vitamin business is a (multi) billion dollar industry and our health has steadily worsened. So what good are dead vitamins? Think it over! Only LIFE begets LIFE! Only LIFE supports LIFE and only live vitamins will save us. Such is GreenLife. —V.E. Irons

V. Earl Irons was born in 1895 and grew to become one of the original pioneers of the natural foods industry. He is most famous for his belief in bowel cleansing and was first to use psyllium seed and bentonite as intestinal cleansing agents. The company he founded in 1946, still manufactures his Sonné's, Vit-Ra-Tox and Springreen cleansing products. Like many of the heros discussed in this chapter, Irons suffered a debilitating disease that changed his life. At age 40, he was stricken with a severe arthritic disease known as ankylosing spondylitis. In desperation, he turned to natural solutions. He attributed his regime of detoxification with the successful restoration of his health and proceeded to promote these principles for the rest of his life.

Irons was no farm boy. He graduated from Yale in 1919 and lived in Boston most of his life. But Dr. Royal Lee, the founder of the famous Standard Process supplement company, brought Irons to Kansas City and introduced him to the wondrous nutritional and detoxifying benefits of grass. He started out by buying grass tablets and powder from Cerophyl and added them to his line of intestinal cleansers which he established in 1946. Impressed with the power of grass foods, he opened an office in the very same building in Kansas City where Cerophyl was headquartered and Schnabel and Kohler worked. Eventually, he bought out one of Schnabel's former farms and started growing and packaging his own cereal grass products. He named it *GreenLife*.

Irons vacuum dried his grass juice into powder at a low temperature to avoid the use of any "carrier" additives. He believed in the "juice" powder instead of dehydrated whole grass because the cellulose is indigestible. "We don't have two stomachs like cows." After beating a debilitating mid-life disease, Irons ended up starting a second family at the age of 72, fathered his last child at 80, and at 86, moved his entire grass farm which included the equivalent of 114 freight cars worth of tractors, trucks and equipment. He died in 1993, just before his 99th birthday. His company, one of the oldest in the nutrition industry, still has offices in Kansas City and is the oldest existing grass products producer in the world. *(See: Companies)*

Verdict of the Guinea Pig Jury [6]

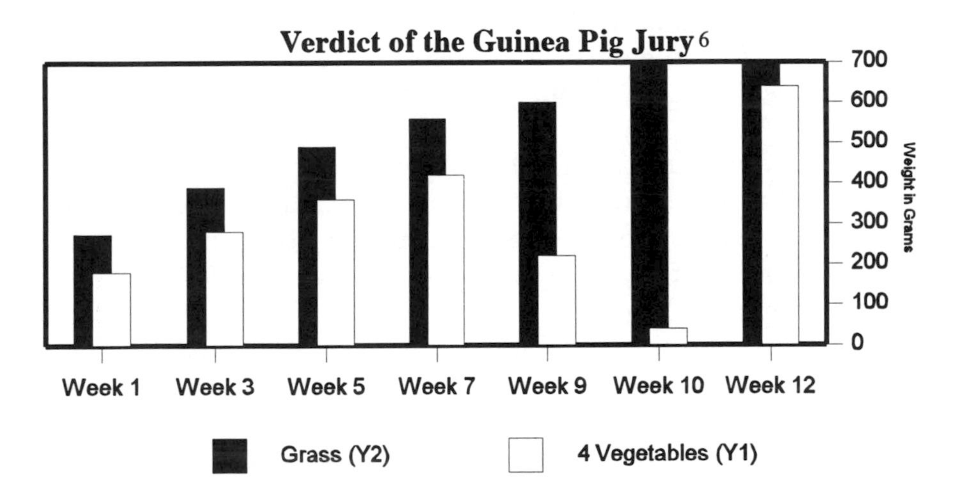

[6] N*otice that even when four of our best vegetables (spinach, carrot, cabbage and lettuce) were fed the guinea pigs, the growth was erratic and the pigs started to fail rapidly after the seventh week. To save his life, dehydrated cereal grass was added to his diet (11th week) and in less than a week, the trend was reversed and a consistent and rapid growth took place. The experiments prove conclusively that while other foods are good, the grasses alone are the complete food and contain all the elements needed to support life.* —V.E. Irons

Photo by Michael Parman

The Story of Ann Wigmore

I see a world without sickness...a world in complete harmony and in perfect physical mental, and spiritual balance by following nature's laws of cause and effect.
—Ann Wigmore

Despite our historic love affair with grains and the worldwide abundance of grasses, no culture has ever cultivated the drinking of grass juice until the latter half of the twentieth century. The popular consumption of fresh squeezed wheatgrass juice is due largely to the efforts and genius of one woman, Ann Wigmore.

Sickness and Adversity Lead to Innovation

Born in Lithuania in 1909, Ann Wigmore had a rough beginning. She started out being born prematurely. Then she was abandoned because she was a sickly baby and a burden to her parents who were seeking a new life in America. Her grandmother, a self-taught naturalist, rescued and restored Ann to normal health. She learned a lot about healing by watching her grandmother heal wounded WWI soldiers with herbs and weeds. But at age 16, she still had no schooling and couldn't even write her name! At the urging of her grandmother, she left for America to get a proper education and reunite with her estranged parents. She wanted to do well in America and adopted the American lifestyle, including a typical American diet. Eventually this resulted in colon cancer. Then a terrible automobile accident shattered both her legs. Gangrene set in and the doctors recommended amputation. She refused and even against her own father's wishes, was sent home. *My homecoming was not a happy one. Neither my father nor my mother would come near me, and only with the help of my uncle would I find something for breakfast.*[7]

Ann knew there was a better way. She returned to her previous "peasant" diet of vegetables, grains, seeds and greens and restored her health by doing what she saw her grandmother do. She picked wild weeds and greens and applied them to her feet. This was not the desperation of a diseased mind, but the result of her grandmother's teaching about the healing powers of grasses and weeds. Ann developed a ravenous appetite for anything green. She nibbled on grass and sucked out its juice. She sat for hours in the warm summer sun watching the greenish blue ring, the "creeping death" rise up her legs. Winter was approaching. There would be no more fresh grass. What to do?

> I *asked God for direction. He supplied an exciting solution. The use of grains to grow greens right in the kitchen*!

One day to her surprise, the little white dog who gave her so much needed love and affection, started to lick her legs. This was the one part of her body he never touched! "My first thought was for the animal's safety, I impulsively raised my arm to move it away, when the injunction of my grandmother came to mind: *Instinct-guided creatures, left to themselves, do not make mistakes."*

That puppy was the first indicator of Ann's recovery. Rest, sun, wild herbs, weeds and kitchen–grown grasses rejuvenated Ann's health. She knew she was not going to die. The doctors informed her father that she was apparently out of danger.*"This infuriated him, because he couldn't accept that he was wrong in his decision to have my feet removed."* It was several months before her feet were completely healed and she returned to the hospital for an examination. The doctors *"made no comment when X-ray films showed that the bones had knitted firmly,"* said Ann. Years later, Ann Wigmore ran in the Boston Marathon.

Ann tested her indoor grasses on her animal friends. Wheat became her favorite grass because the animals chewed more of it and it was sweet tasting, easy-to-find and inexpensive. To further her studies, she even adopted a sick, cancerous monkey. Ann nursed the monkey back to good health with creative techniques and live food recipes including: sprouted seeds, fermented nut and seed "yoghurt," and rejuvelac, a cultured sprouted wheat drink. These, along with wheatgrass, would later become the cornerstone of her Living Foods Diet.

Though she knew she had stumbled on something very valuable, she had no practical way of making it widely available. Most people would not eat their front lawn no matter how sick they were! There had to be a better way. Then, at a local yard sale, Ann picked up an old cast iron meat grinder on the chance that it might grind the grass. With small modifications, such as the addition of a stainless steel straining sieve, this grinder became the first wheat grass juicer. It was a real innovation because commercial vegetable juicers could not at all manage the ligneous fiber of grass. This made possible her program whereby anyone could grow grass in their kitchens and extract the juice in their homes.

Ann began delivering fresh wheatgrass juice to bedridden, ill and elderly people in her Boston neighborhood. Then in 1958, she turned an old mansion on Commonwealth Avenue in Boston into The *Hippocrates Health Institute*. It was founded on the oft repeated principle of Hippocrates, the Greek father of modern medicine, who along with his hippocratic oath, is often paraphrased as saying: "The body heals itself. The physician is only nature's assistant." Dr. Ann, as she was fondly called after she became a doctor of Divinity, believed that the body can act as its own physician given the proper tools—living foods. *"Living foods for living bodies, dead foods for dead bodies,"* said Ann.

Although Dr. Ann was neither a marketer nor a scientist, Hippocrates Health Institute, situated in the heart of Boston, attracted many celebrities and researchers who helped give testimony to the healing powers of grass. Among these were Dr. Chiu-Nan Lai, Dr G. H. Earp Thomas, Renee Taylor, Dennis Weaver, and Dick Gregory. Even Yoko Ono purchased wheatgrass for juicing in her home. People came from all walks of life and all corners of the world, usually as a "last resort," with the hopes of beating cancer or some other degenerative disease.

> *Vibrating with health and vitality, Ann Wigmore is constantly on the run. Reaching out to all that are open to hear what she has to say, she emphasizes that anyone can live a healthier, happier and more fulfilling life.* —Dennis Weaver, actor,1984.[8]

Ann had two visitors who were enormously helpful in promoting her work. One was Viktoras Kulvinskas (*see p. 35*), who helped Ann establish the Hippocrates Health Living Foods program, and the other was Eydie Mae Hunsberger. Eydie had breast cancer. Her surgeon told her: "You have an 80% chance to live one year and a maximum life expectancy of five years." Eydie chose to have a lumpectomy that removed the cancerous tissue but it had spread. Depressed and frightened, she looked everywhere. She ended up at Dr. Ann's door in 1973. After two weeks on Dr. Ann's Living Foods Program, Eydie was confident she would beat cancer. Two years later, with no trace of cancer in her body, she wrote *"How I conquered Cancer Naturally"*[9] which was very successful and brought many people to Dr. Ann's door.

Dr. Ann has a long list of stories and testimonials from guests who improved their health with the help of wheatgrass. Wheatgrass and its sister triticum barley, have by testimony, helped guests with ailments like high blood pressure, diabetes, obesity, gastritis, stomach ulcers, pancreas and liver troubles, asthma, glaucoma, eczema, skin problems, constipation, hemorrhoids, diverticulitis, colitis, fatigue, female problems, arthritis, athlete's foot, anemia, bad breath/body odor, and burns. In addition, wheat grass has served as a wonderful first aid for red eyes, wax in ears, congested nasal passages, bleeding gums, tooth pain, sore throats, and inflamed mucous membranes.[10 11]

Ann tried to tell the U.S. government about wheatgrass. She even went to Washington. But the political and nutritional climate of the 1970's were stubbornly closed–minded. She had a much better reception abroad. Ann visited some twenty countries and launched living foods programs in India, Sweden, Finland and Canada. Ann learned that sickness disregarded all borders.

> *Let us make a concerted effort to remedy the global problems by correcting the physical and mental imbalance in each of our lives...The industrialization of our society has created an artificial lifestyle in which humans are being led further and further from the basic truths inherent in nature.*[12]

In February of 1994, Ann Wigmore died of smoke inhalation in a middle of the night fire that destroyed the Boston home of the original Hippocrates Institute. She was nearly 85. Although she has passed, her work continues through the efforts of the many lives she touched. Her single clinic in Boston now has offsprings in six locations in the USA alone and others in Australia, Sweden, Finland and India. Her teachings changed the lives of millions. *(See Epilogue.)*

> *My life has been, and continues to be, dedicated to the wellness of all humanity.* —Ann Wigmore, 1909–1994

Viktoras Kulvinskas

Wheatgrass juice is the nectar of rejuvenation, the plasma of youth, the blood of all life. The elements that are missing in your body's cells–especially enzymes, vitamins, hormones, and nucleic acids can be obtained through this daily green sunlight transfusion.
—Rev. Viktoras Kulvinskas, M.S.

He arrived from the Massachusetts Institute of Technology. He was a mathematician and computer consultant at Harvard University. He was young, but not well. He started making some of the hard choices many of us are never able to make. He took his health in his own hands. He quit the University; turned his back on a successful career and all the security that came with it. He walked into Dr. Ann's mansion, resurrected his health and in the process became a Gabriel for Dr. Ann and an archangel for the living foods movement.

Viktoras was a marketing bonanza for Dr. Ann's teachings. He was the best missionary anyone could wish. In his second year there, he took Dr. Ann's book, which at the time was a magazine entitled *Be Your Own Doctor,* on a Shiloh Farms delivery truck and showed it to seventy health food stores. His "gorilla" marketing doubled her clinic's attendance from 3–6 to 6–12 students. Later, he came out with his best selling new age manual, *Survival in the 21st Century*. That inspired work, an amalgam of art, poetry, spirituality, nutrition, and gardening served to inspire a whole generation. It quickly became a classic. This author was one of the "children" transfixed by its spell. Truly a 1970's book, it fit right in with the "tune in, turn on, and drop out" theme of the era. Only, it had nothing to do with drugs and everything to do with personal and planetary healing. It was rebellious against the S.A.D. (Standard American Diet) without any words of malice. It condemned M.D. (Medical Dogma) without harshness or anger. And it all came from a very "grass roots" (no pun intended) printing effort. The love and strength of his message quickly developed a following and Viktoras became an apostle of the new age.

Survival Into the Twenty First Century was also one of the first successes of the small press. Viktoras called his publishing company O'Mango'd Press. This name illustrates Viktoras' creativity and joyful spirit. It can be interpreted as being either about mangos or Man/God. The first interpretation is perfectly suited for a raw foodist press; the second is ideal for a spiritual philosophical press. His press is both, and in this one moniker he articulates the dual theme of his work.

Today, Viktoras lives in beautiful Hot Springs, Arkansas. As this book goes to press, he is in his fifty–eighth year and still taking on major projects including the new *All Life Sanctuary* retreat center *(See Resources)*. He still grows wheatgrass and still believes in raw, chlorophyll rich foods. He is a key promoter for the use of blue-green algae, another super-chlorophyll rich food. On a typical day, Viktoras works longer, sleeps fewer hours and tires less than his younger colleagues.

> *If only they knew, the use of grass is the most revolutionary concept introduced into the diet of society...In therapeutic amounts, wheatgrass internalizes a maximum of green chlorophyl and enzyme rich liquid food, to detoxify the body by increasing the elimination of hardened mucus, crystallized acids and solidified, decaying fecal matter. Its high enzyme content helps to dissolve tumors. It is the fastest surest way to eliminate internal waste and provide an optimum nutritional environment, so that the cosmic cell consciousness can rebuild your body.*
>
> —Rev. Viktoras Kulvinskas, MS., *Survival Into the 21st Century*

Yoshihide Hagiwara

It was clear to me...that the leaves of the cereal grasses provide the nearest thing this planet offers to the perfect food...It is my belief that the steady depletion of that natural green power in the human diet, and its displacement by other nutrients of questionable value, constitutes the most serious threat of all to good health. —Yoshihide Hagiwara

About the same time Ann Wigmore was experimenting with wheatgrass juice, a pharmacist and medical doctor on the other side of the world started exploring the use of barley grass. Both barley and wheat are members of the triticum grain family. They are brother and sister seeds. It takes a close look to tell them apart. It is easier to see the differences between Wigmore and Hagiwara. Their nationalities and cultural backgrounds would be enough. But Wigmore had no marketing, science, business or financial background. Hagiwara excelled in all of these. In spite of these differences, there were some dramatic commonalities. Both had troubled childhoods and dramatic health crises. But most importantly, both were fervent believers in the power of grass.

Yoshihide Hagiwara got off to a tough start. His parents were unable to care for him and gave him away, just like Wigmore. And when his foster-mother developed cancer, she put him in an orphanage. Nevertheless, young Hagiwara had the courage and strength of character to step out on his own. He joined the Naval Academy and after he was discharged, enrolled in the University to study pharmacology. Upon graduating in 1949, he opened up a pharmacy and started inventing different formulas for his customers and neighbors, including preparations for the skin, feet and hair. Some of these are still sold in Japan today. Later, he would go back to school, this time to become a Doctor of medicine.

In 1952, he moved to Osaka and opened a drug manufacturing company named Yamashiro Pharmaceutical Co. Ltd. There he developed and patented over 200 medications and amassed a staff of over 700. He had created the largest drug company in Japan! Hagiwara loved doing research so much, he never stopped working in the lab. But long hours, bad eating and sleeping habits, stress, and working with dangerous chemicals such as mercury, took its toll on his health. At age 38, his hair turned gray, his teeth fell out, his mind clouded up and he could no longer manage a large company. He was forced into bankruptcy.

Hagiwara, ever the fighter, mustered the strength to resurrect his health. Like Ann Wigmore, he was influenced by Hippocrates—"A disease is to be cured by man's own powers." He also studied Shin-Huang-ti, the ancient Eastern physician who compiled the fundamentals of Chinese medicine. He made a determined effort and completely transformed his diet using many Chinese herbs and fresh greens. His health returned and once again he turned his energies to finding similar solutions for others. *"I began a personal search for a food which would promote good health by vitalizing the body's own power of healing."*

Hagiwara's unique position was that he had the business acumen, resources and mettle to deliver a pioneering product to the modern marketplace. He was a doctor, a pharmacist, an inventor, and an experienced businessman. Perhaps the secret to Hagiwara's success, and in striking contrast to Wigmore, is that he knew in order for a product to succeed, it had to be palatable and convenient to use.

> *To fit into the modern world, it had to be easy to purchase, easy to use, and stable over a long period of time....What I wanted was something with a modern look and form, an ideal 'fast food'.*[13]

While Wigmore's approach was to stimulate people to "be their own doctor," and grow their own food, Hagiwara wanted to give them a "fast food." His dream was to make the healing powers of green vegetables as easy to have as instant coffee. He believed that the juice of green vegetables was the finest source of nutrition. Ever the scientist, he analyzed 150 edible green plants including chickweeds, asters, pigweeds, clovers, kudzu, peas and acacias. And just like Charles Schnabel and George Kohler before him, he found that cereal grasses—barley, rye, wheat, and oats, had the most remarkable quantities of active ingredients.

Even before he turned his efforts exclusively to barley grass, Hagiwara was testing drying equipment on herbal teas. He figured if he could dehydrate Chinese herbal tea and maintain the medicinal benefits, then consumers could use these wonderful herbs as easily as they have instant coffee. Hagiwara the inventor successfully developed and patented a spray dryer that in only 3 seconds turned Chinese herbal tea into powder. He borrowed money from his family and friends and in 1971 established Japan Natural Foods Co. Ltd. They started out marketing high quality, powdered Chinese herbs.

Hagiwara wanted to make the benefits of green foods as easy to get as instant coffee.

Next, he turned his efforts to the cereal grasses but things didn't go smoothly. Customers were not getting results. He realized that while coffee is spray dried at 130°F, which adds a roasted aroma, this was too hot for the active ingredients in barley leaves. He refined the technique to dry at room temperatures, but the product turned brown. There were also problems with the harvesting and growing. It turned brown just sitting in the truck! He tried freezing the leaves, but that inactivated the medicinal ingredients. Like V.E. Irons in 1951, Hagiwara realized the best solution was the most expensive one—build the production facility near the fields. Although this raised production costs, he was undaunted. He was driven to perfect the product at any cost and refused to compromise. He knew that the plants needed to be harvested when they were only 8 to 10 inches tall since that is when they contain the maximum amounts of active medicinal ingredients. But this reduced the crop yield by one third over harvesting at 16–18 inches. Through trial and tribulation, Hagiwara ultimately achieved a dried barley green juice powder that maintained the nutritional integrity of the fresh plant.

In 1980, ten years after its introduction in Japan, Hagiwara brought his green barley leaf powder to the USA. In 1990, he made a major investment in American health and opened a $21 million dollar growing and processing facility in California. Hagiwara is the best kind of entrepreneur—one who puts his profits back into research. In 1981, he established the Hagiwara Institute of Health, which is today, the major source of nutritional information about the medicinal properties of grasses. Dr. Hagiwara's son, Hideaki Hagiwara Ph.D carrying on his father's love of science, is its director. Numerous scientific studies evaluating the medicinal properties of barley leaf juice extract have been completed in concert with scientists at the University of Tokyo, the University of California at Davis and George Washington University, Washington DC. *(See chapter: Research)*. This research of grasses in relation to human health and nutrition is the most extensive to date and arguably is the most important legacy Yoshihide Hagiwara leaves to us.

> *What I am suggesting is that these 'medical' tools have had the unfortunate effect of corrupting attitudes toward health. This has gone so far, I fear that now many people believe in medical and technological guarantees of good health, while little by little they give up their personal responsibility for the good health of their bodies.*[14] —Yoshihide Hagiwara

In 1994, Japan's Minister of Health and Welfare awarded Yoshihide Hagiwara the Drug and Medical Meritorious Service Award for his development of the extraction techniques used to manufacture botanical products for people and pets. While Hagiwara is clearly a pioneer in the natural foods industry, he should also be recognized for bridging the world of conventional medicine and pharmacology with natural products. He delivers scientific respect, recognition and credibility to the world of natural foods that few others have accomplished.

Accomplishments of Yoshihide Hagiwara

Inventor, Pharmacist, Entrepreneur, Medical Doctor. Founded, established and achieved the following in his lifetime:

- 1949. Founded Hagiwara Pharmacy
- 1952. Yamashiro Pharmaceutical Co., Ltd.
- 1968. Japan Pharmaceutical Development Co., Ltd.
- 1969. Hagiwara Physical and Chemical Laboratory
- 1971. Japan Natural Foods Co., Ltd.
- 1976. Green and Health Association
- 1978. Vice Pres. of Institute of Manufacturing all Chinese Medicine, a national institute for manufacturers.
- 1979. Director of Research Institute of Health Foods of Japan.
- 1980. Counselor of the Institute of Drugs of Osaka.
- 1981. Counselor, Association of Medicine Manufacturing Industry of Japan.
- 1981. Hagiwara Institute of Health for cancer research.
- 1987. Award from Japan's Science and Technology Agency.
- 1994. Drug and Medical Meritorious Service Award presented by Japan's Minister of Health.

Nutrition

Grass does many things in animal nutrition which cannot be accounted for by its known vitamin content.
—Dr. Charles F. Schnabel

Grass Is More than the Sum of its Parts

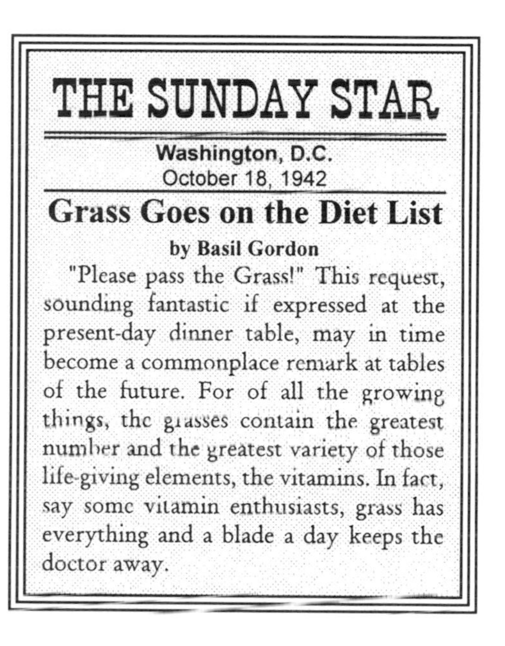

THE SUNDAY STAR

Washington, D.C.
October 18, 1942

Grass Goes on the Diet List

by Basil Gordon

"Please pass the Grass!" This request, sounding fantastic if expressed at the present-day dinner table, may in time become a commonplace remark at tables of the future. For of all the growing things, the grasses contain the greatest number and the greatest variety of those life-giving elements, the vitamins. In fact, say some vitamin enthusiasts, grass has everything and a blade a day keeps the doctor away.

The vitamin supplement industry is booming. Consumers are anxious to take nutritional products as an alternative to drugs. This is a welcome trend, but where do supplements come from? Plants have been the source of our medicines for thousands of years*(See p. 175)*. Our modern drugs are largely synthetic replicas of nutritional factors found in plants. Our earliest drugs—aspirin, penicillin, quinine—all came from nature. Since our food is our medicine and our medicine is our food, as the father of medicine Hippocrates is famous for saying, then the best medicine is a whole, natural food.

Even the earliest research of the 1930's identified the young cereal grasses as complete foods. All known nutrients were found including other unidentified ones called "grass juice factors." Today, they would be labeled phytochemicals. Guinea pigs lose weight and weaken on a diet of mixed vegetables but thrive on grasses. Mega doses of vitamin supplements can function like medicines to stimulate or enhance biological function. But nutrients don't live in a vacuum. They co-exist with numerous other factors that enhance and enable their function. Once isolated, they may not work as well. Albert Szent-Györgyi, the Nobel prize biochemist who discovered Vitamin C, found the complex structure of 'C' in peppers was more effective than the chemically isolated 'C.' When we start adding isolated vitamins of different kinds to our diet, it is a case of man trying to replicate nature. Nutrition by the numbers. How many milligrams of calcium in relation to potassium is the perfect balance? A whole food is a complex bundle of thousands of chemicals. Squeezing out isolated nutrient fractions dissipates its magic. Grass contains hundreds of vitamins, minerals, enzymes, amino acids, phytochemicals, anti-oxidants, cellular RNA and DNA all in concentrated form. Like grass, spirulina, chlorella, blue green algae, and bee pollen are a few other wonderful natural foods containing a broad spectrum of concentrated nutrients. These foods give our bodies the raw materials from which it manufactures what it needs and balances its own chemistry. The philosophy of this book is that where basic nutrition is concerned, concentrated whole foods make the best supplements.

> *It is folly to dose ourselves with one or two vitamins when we know nothing of their relationships to fifty other food factors.*
> —Charles F. Schnabel

Nutrition Depends on Where, When and How

It all starts in the soil. The earth feeds the plants and the plants feed us. When you walk into your natural food store to purchase grass, it may have come from the high terrain of Utah, the plains of Kansas, the Pacific Coast of California or the basement grow room of your local grower. They can't all be nutritionally identical. True, they may all be organically grown, but that still leaves many, many variables. There is no equality in nature. Regardless of whether you are growing strawberries, tomatoes or grass, the nutritional analysis will vary from state to state, grower to grower and farm to farm. From soil, to water, to weather, disorder is the fundamental law of nature. Was the season too hot or too cold? Too wet or too dry? Growing times vary from 60 days to 200 days in the field and 8–14 days in the greenhouse. Some tray grass growers can eliminate a few variables with climate controlled greenhouses and automated irrigation. But processing is yet another influence on nutrition. Is your grass fresh or frozen, freeze dried or spray dried. Drum dried or evaporated? Was the whole leaf ground and powdered or was the grass juiced and dried? What was the method of juicing? How is it bottled? In glass or plastic? Is it colored or clear? Is it nitrogen flushed? Even two samples of grass squeezed fresh in front of you look and taste slightly different depending on who did the growing, where it was grown, what type of soil, seed, temperature, water, harvest time, etc.

True, there are many variables, but don't be dismayed. Grass is a food and these variables exist with every fruit and vegetable you purchase. Of course, you can always grow your own. You can be in charge of these variables, but you won't be able to control them all. Are you ready to invest the time and materials? Are you sure you can do a better job than the professionals? Can you be sure that you will produce a nutritionally superior grass? (For more, see chapters: *The Companies* and *Grow It.*)

Harvest at the Peak of Nutrition–The Jointing Theory

In 1930, Charles Schnabel fed his sick chickens grass. They completely recovered and furthermore achieved phenomenal fertility. As an agricultural chemist, he was able to figure out why. The young oat grass had something special about it that wasn't in the mature grass. "My only clue was that the hens refused to eat the older grass."[1] Schnabel repeated his experiments the following summer and isolated the exact moment of its maximum nutrition. The differences in nutritional content were so remarkable, depending on the time of harvest, that Schnabel eventually received a U.S. patent for his discovery. Essentially it is this: the plant undergoes rapid growth with increasing nutritional manifestation until maturity when it switches gears from vegetative growth to reproductive growth. Just prior to this changeover, the plant is at its vegetative peak. This transition is known as jointing. Technically, it is when the ovul of grain moves up the staff from the root. It only takes a few days. After jointing, the nutritional counts drop radically. The plant sends all its nourishment into the developing seed. The grass grows quickly, and starts blooming and producing grain. Although harvesting a week later would produce twice the amount per acre, the nutritional level would be dramatically inferior.

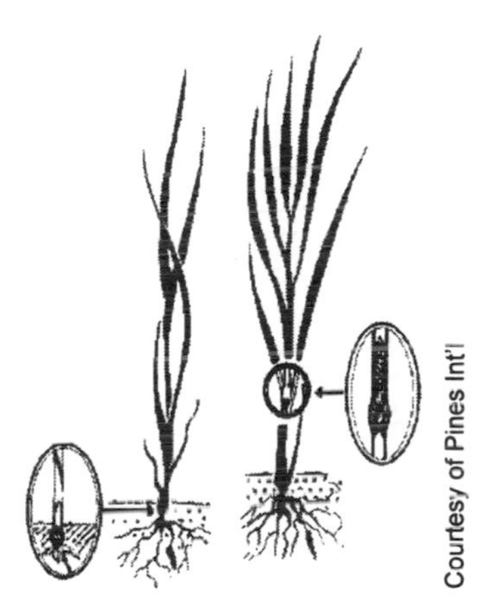
Courtesy of Pines Int'l

> ...*It should be obvious that the phenomenal feeding value of good grass at the jointing stage is a transient quality. If it is not grazed or cut and preserved within a few days of the jointing stage, its value is lost forever because the vitamins associated with photosynthesis are not transferred to the grain or returned*

to the soil. They are used up by the grass plant during the reproductive stage of growth. It is this transient nature of grass quality at the jointing stage that has made the secret of grass so elusive. The irony of it is that our "dumb" animals seem to have known the secret all the time.—Charles F. Schnabel[2]

Because this discovery had tremendous influence on the livestock feed industry and secondarily on human nutrition (via milk and meat), there was much interest in verifying it. In 1935 Phillips and Goss reported a field of barley contained 38.8% protein on the 21st day of growth (the jointing stage), 12.2% protein on the 49th day (the bloom stage) and 3.8% protein on the 86th day (the mature stage).[3] Also that year George Kohler, the most active grass researcher of his time, reported that most of the vitamins in grass reach a peak per gram of dry matter at or near the jointing stage and that roughly parallels the protein content. This peak may be twice as high on the day of jointing as it is a week before or a week after.

Indoor Grass vs. Outdoor Grass

Tray grown grass, grown indoors or in greenhouses, never achieves the jointing stage. However, this grass is always harvested on its way to peak. Protein counts are typically an excellent 40–45% on a dry weight basis. Fresh squeezed tray-grown grass juice is two percent protein. But, since the juice is 95% water, this represents 40% of the solid content. Indoor grown grass, under warm conditions, has achieved jointing as early as the 18th day. Since tray growers typically harvest between 8 and 14 days, they are coming close. Considering that they can get close to jointing and that the grass is fresh squeezed and drunk immediately, this approach has all the nutritional advantages of a fresh vegetable close to its prime.

On the other hand, because it grows so quickly under warm conditions, the plant has a relatively high level of simple sugars. You can taste the sweetness. In the outdoor grass, some of that sugar is converted into more complex carbohydrates and vitamins. The accelerated growth causes the indoor plant to put most of its energy into growing leaves rather than roots. Thus, very few minerals can be absorbed from the roots and utilized to produce more complex nutrients. The warm temperatures and low air circulation also cause growing problems. Mold is the scourge of all indoor growers and that is because grass likes it cool. Although grasses grow all over the planet, they like cool places like the heartland of America (the grass industry grew up in the 1930's in Kansas City, Missouri) where amber waves of grain blow in the cool breeze and snow blankets the winter wheat. Although indoor grass is grown in every state in the union, professional growers in hot climes like Florida and southern California use fans, shades and even climate controlled greenhouses to keep mold in check. But kitchens and greenhouses cannot match the great outdoors. The soil in trays is only 1–2 inches deep and the roots are barely mature enough to assimilate nutrients no matter how good the soil. In contrast, outdoor grass develops deep roots, pulling up minerals and manufacturing vitamins over 60–200 days of slow growth. The seeds are widely sown about three inches apart, allowing them to 'stool out' or form additional leaves (culms) from the roots. Sun beats down on the field crop for 4–8 weeks (longer for winter wheat). Such exposure and slow growth in the cool fall or spring turns the grass into a solar collector, storing high concentrations of energy in its leaves. This provides a full spectrum of chlorophyll, trace minerals and micro-nutrients. This is how wheat was designed to grow. Plastic greenhouse roofs and kitchen windows filter out some light waves including the important ultraviolet. This may deprive us of receiving the full spectrum of nutrients that long hours of unimpeded sunlight can offer. Nevertheless, both grasses grow to approximately 7–10 inches tall. The extra days of growth outdoors are spent building roots and forming additional culms.

In nature, the purpose of the plant is to produce grain. Once it has accumulated enough solar energy/nutrition, it generates the wheat berries which we collect and grind into flour. By harvesting the plant just before it joints, we are taking advantage of this "vegetable" at its nutritional prime. Although you can grow it indoors and come close, the finest wheat and barley grass is grown outdoors. The best of both styles would be to grow your grass in nature and fresh squeeze the juice indoors. No matter which grass you choose, you will be getting good grass. The quality of the grass produced across the industry is high. Grass growers are dedicated types, regardless of whether they have thousands of acres or a basement grow room. No one grows grass to get rich quick. Grass growers are devoted to providing a product that will help people revitalize their health.

- Life Sustaining Complete Food
- Economical Nutrients
- Concentrated Overall Nutrition
- Vitamins C, E & K and Oxygen
- Beta Carotene (A) & Chlorophyll
- Antioxidants and Detoxification
- Quick Assimilation
- Enzymatically Active
- Immune Stimulation and Defense

What's In Grass

All vegetables are food factories. Each excels in different nutrient areas. Because whole dehydrated grass is 25% protein (meat has 17% and eggs have 12%), we could define grasses as protein foods. But there are other foods like algaes with higher protein than grass. So why should we even bother with the lower protein foods? Nutrition is not a weight lifting contest. It's not about more protein, but more equilibrium. It's not quantity, but quality and the most important quality factors are diversity and balance. Grass is a balanced food containing a broad spectrum of high quality vegetable nutrition. Grass is a rainbow food and at the end of the rainbow is indeed a pot of golden health. Grass does not zoom you there with speed, but blows you there gently on the magic carpet of endurance and delivers you with longevity.

> *The grain of grass contains all the elements of which the body is composed including revitalizing and rebuilding materials, force producers for energy and also the eliminators of waste acids. Grass and sprouts are perfect foods.*—Dr. Ann Wigmore

Grasses are a complete life sustaining food. Based on the animal studies *(see Research)*, if you had to choose one food for survival, it ought to be grass. After all, grasses are the primary food for domestic and wild grazing animals and pretty large ones at that—cows, horses, goats, sheep, buffalo, deer, giraffes. The guinea pigs, chickens, rats and other laboratory animals were losing weight on our finest vegetables but quickly reversed their downward trends when switched to grass. (See chapters: *Pioneers, Research.*) These dumb animals have long known the truth about grass. We are just coming around now because we have the technological means to compensate for our inability to digest the grass directly. Instead of four compartments in our stomachs, we have juicing machines and micro-fine powdering equipment.

Fifteen pounds of wheat grass is equivalent to 350 pounds of the choicest vegetables. —Charles F. Schnabel

Blue-green algae, chlorella and spirulina are wonderful and important superfoods that surpass grasses in certain nutrient categories and should be in our diet. But they are more expensive to cultivate than grasses and their cost per nutrient is high. Man has evolved from the land and historically, land grown foods are best suited for land based animals. Charles Schnabel was so impressed with the inexpensive cost of vitamins from grass, he created a "Yard-Stick" for measuring the value of foods in terms of pennies per nutrient. According to Schnabel, "only one-half ounce of 40% protein dehydrated grass would supply 18,600 units of vitamin A, 113 milligrams of vitamin C and 5.7 grams of the best quality protein in the world. This is more vitamin A than supplied by the entire 84.5 ounces of food in the (USDA recommended) optimum diet and more vitamin C than is supplied by the entire 30 ounces of fruits and vegetables."

Nutrients in Grass

Amino Acids: Tryptophan, Glutamic Acid, Alanine, Methionine, Arginine, Lysine, Aspartic Acid, Cystine, Glycine, Histidine, Isoleucine, Leucine, Phenylalanine, Proline, Serine, Threonine, Tyrosine, Valine.

Enzymes: (Over 80 have been identified) Peroxidase, Phosphatase, Catalase, Cytochrome Oxidase, DNase, RnaseSuperoxide, Hexokinase, Malic dehydrogenase, Nitrate reductase, Nitrogen oxyreductase, Fatty Acid Oxidase, Phosolipase, Polyphenoloxidase, Dismutase, Transhydrogenase. Phytochemicals: Chlorophyll, Bioflavonoids

Vitamins: Vitamin C, vit. E (succinate), Beta-carotene (vit.A) Biotin, Choline, Folic Acid, B1-Thiamine, B2-Riboflavin, B3-Niacin, B6-Pantothenic Acid, vit. K

Minerals & Trace Minerals: Zinc, Selenium, Phosphorus, Potassium, Calcium, Boron, Chloride, Chromium, Cobalt, Copper, Iodine, Iron, Magnesium, Manganese, Nickel, Sodium, Sulfur. (These are the primary ones, there are many more.)

Fatty Acids (essential): Linolenic Acid, Linoleic Acid.

Grass is a wonderfully balanced source of nutrients. Excellent for all minerals major and minor, it is especially high in calcium, magnesium, manganese, phosphorus, and potassium, as well as trace minerals such as zinc and selenium. All are important for cardiovascular and immune system function. In the B-vitamin department, grass has them all, including the crucial biotin, folic acid, pantothenic acid, an abundance of choline (lecithin) and is a vegetable source of B-12. Protein is 2% in fresh wheatgrass juice and up to 45% in barley grass juice powder. An egg, the symbol of fertility long considered the perfect protein, is 42%protein (dried). Protein in grass is in the form of poly-peptides—simpler, shorter chains of amino acids—that enable faster, more efficient assimilation into the bloodstream and tissues. Grass includes at least 20 amino acids both essential and non-essential. Its spectrum of vitamins is so broad, that in 1939, dehydrated wheat grass was actually accepted by the American Medical Association as a natural vitamin food. *(See p. 28)*

The Magic of Chlorophyll

Whenever anyone talks about the healing powers of grass, they mention chlorophyll first. Grasses, along with alfalfa and algaes, are the richest sources of chlorophyll on the planet. No surprise here. One third of the planet is covered with grass including the one inch tundra above the Arctic Circle. Chlorophyll and the omnipresence of grass are essential to life on the planet. Green plant cells are the only ones capable of absorbing energy directly from the sun. The primeval energy for all life is thus light. Sunlight radiation is absorbed by plants and secondarily by humans and animals. If the energy output from the sun were to cease, the basic vital functions of all living organisms would gradually slow down, and eventually life on Earth would become extinct. It takes eight minutes for a photon of light to travel the ninety-three million miles from the sun to the Earth's surface. A green plant needs only a few seconds to capture that energy, process it, and store it in the form of a chemical—chlorophyll. This process of converting light into energy is called photosynthesis.

> *All forms of life, on land and in the sea, even when they feed on each other, are parasites, depending ultimately on plant life. Your body, its flesh and its organs are made up largely of protein coming to you directly from food plants or the flesh of plant eating animals. You personally, exist only because of chlorophyll.*—Dr. T. M. Rudolph[4]

PHOTOSYNTHESIS CREATES LIFE

Cells in leaves do miraculous things. Water enters through the veins and CO^2 through the stomata at the bottom. Photons of light from sunshine are captured in cells called chloroplasts. As the chloroplasts absorb the light, their electrons become excited. They are literally dancing in the sunshine. They are charged with energy which is stored as ATP (adenosine triphosphate). The ATP then reduces carbon dioxide and water to oxygen and carbohydrates. The oxygen exits the leaf and fills the atmosphere with fresh air. The carbohydrates remain as food. Were chemists able to duplicate photosynthesis by artificial means, we would have the use of the sun—solar energy—for all our energy needs.

Famous research scientist E. Bircher called chlorophyll "concentrated sun power" and reported that it "increases the functions of the heart, affects the vascular system, the intestines, the uterus, and the lungs. It raises the basic nitrogen exchange and is therefore a tonic which, considering its stimulating properties, cannot be compared with any other."[5]

One of the reasons chlorophyll is so effective is its similarity to hemin. Hemin is part of "hemoglobin," the protein portion of human blood that carries oxygen. Studies as long ago as 1911 show that the molecules of hemin and chlorophyll are surprisingly alike. The primary distinction is that chlorophyll is bound by an atom of magnesium and hemin is bound by iron. Experiments proved that severely anemic rabbits make a rapid return to a normal blood count once chlorophyll is administered.[6] The body seems to be able to substitute iron and rebuild the blood, virtually giving the anemic patient a blood transfusion.

Chlorophyll has long been famous for its ability to heal infected and ulcerated wounds. Studies prove that "tissue cell activity and its normal regrowth are definitely increased by using chlorophyll."[7] It is an important medicine for healing bleeding gums, canker sores, trench mouth, pyorrhea, gingivitis, even sore throat. Chlorophyll has the unique ability to be absorbed directly through the mucous membranes, especially those of the nose, throat, and digestive tract. It makes a great mouth wash and

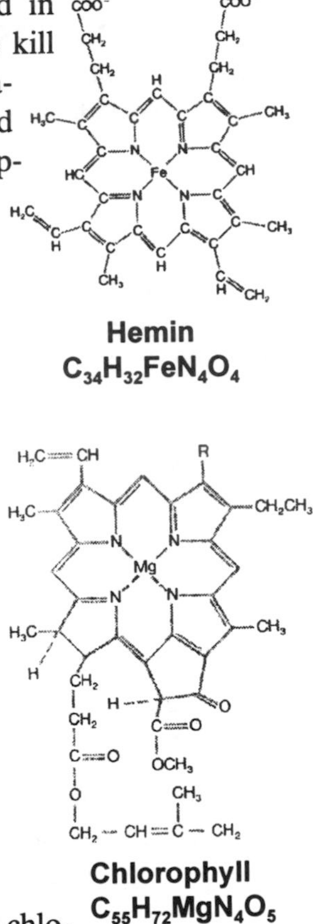

Hemin
$C_{34}H_{32}FeN_4O_4$

Chlorophyll
$C_{55}H_{72}MgN_4O_5$

an excellent dentifrice, especially when used in powder form. Chlorophyll's unique ability to kill anaerobic–odor producing bacteria–is the reason it covers up the smell of garlic, fights bad breath, body odor, and acts as a general antiseptic. These bacteria live without air and are destroyed by chlorophyll's oxygen producing agents. Dr. Otto Warburg, the 1931 Nobel prize winner for physiology and medicine, concluded that oxygen deprivation was a major cause of cancer. At least one alternative cancer therapy today bombards tumors with ozone—highly active oxygen. Unlike many drugs, chlorophyll has never been found to be toxic at any dose. And not one of the 9,000 species of grasses that cover our planet is poisonous.

Chlorophyll may also provide us with protection from low level X-ray radiation from hospital equipment, televisions, computer screens, transmitters and microwaves. No area is totally radiation free. Experiments on guinea pigs in the 1950's demonstrated that radiation poisoned guinea pigs recovered when chlorophyll rich vegetables were added to their diet.[8] The U.S. Army repeated this experiment with broccoli and alfalfa and got the same results.[9]

Grass–A Cornucopia of Important Nutrients

But, chlorophyll is only one of the important pigments in grass. There are other pigments such as carotenoids—alpha-carotene and the famous beta-carotene xanthophylls and zeaxanthin to name a few. There is an abundance of these phytonutrient pigments in grass. Unfortunately, you can't see them because just as with the beautiful autumn leaves, the green chlorophyll overpowers the other pigments. There are up to 18,000 units of beta-carotene per ounce of grass. This pre-cursor of vitamin A has significant immune enhancing properties including the promotion of T-cells. High levels of this anti-oxidizing nutrient are associated with reduced cancer risk and cardiovascular disease.

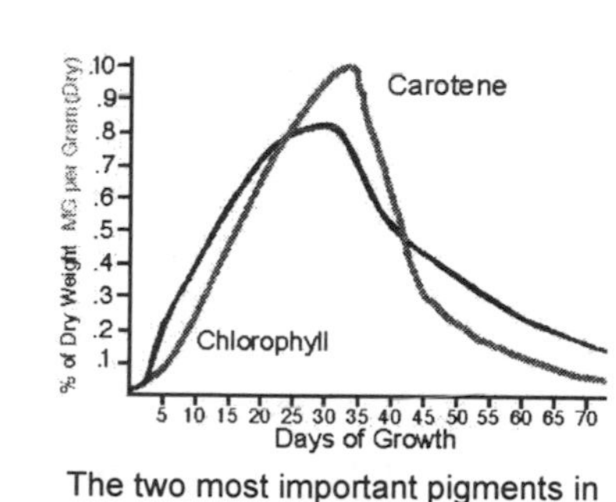

The two most important pigments in grass—chlorophyll and carotene peaking at the jointing stage. Other nutrients follow similar patterns.

Another important vitamin and antioxidant abundant in grass is vitamin E. Grasses have a water soluble form of E called a-tocopherol succinate which stimulates the production of T-cells, antibodies, interleukin2 and interferon among its many immune system functions. This form of vitamin E is very effective in suppressing the growth of cancer cells in vitro. In addition, it has the ability to increase production of prolactin and growth hormone in the pituitary gland. *(See p. 62)* Grasses are also abundant sources of quality vitamin K, the blood clotting vitamin. The body needs it to form the enzyme prothrombin which creates fibrin that clots blood. It also acts as an antidote for certain poisons.

Dr. T. Shibamoto of the University of California, discovered a powerful new antioxidant in barley grass called 2"-*0*-GIV. This new isoflavonoid is both soluble in water and fats and is highly stable. This means it is capable of permeating both the fat and aqueous cell membranes in order to fully protect the cell from the damaging effects of oxidation. According to Shibamoto, 2"-*0*-GIV is more potent than vitamins E and C but when taken together, the effects are profound. Barley grass has all three in good quantity. Tests have shown it is a preventative for arteriosclerosis and is just as effective as the prescription drug Probucol for this disease, without any side effects. *(See p. 64)*

Barley and wheat grass are both abundant, inexpensive sources of superoxide dismutase (SOD). This is a powerful antioxidant and anti-aging enzyme. SOD is a proven anti-inflammatory for arthritis, edema, gout, bursitis, etc. Dr. K. Kubota of the Science University of Tokyo found two glyco-proteins D1G1 and P4D1 which work alongside with SOD but are more heat stable. All three have anti-inflammatory action that is superior to the much touted aspirin.

Dr. Hagiwara tests SOD activity in barley grass as a yard-stick for measuring overall enzyme activity. If the heat sensitive SOD is active, so are all the other 80+ enzymes in grass. When most people think of enzymes they think of digestive enzymes such as lactase, amylase, protease and lipase. But these do not alone digest all foods and digestion is not the only function of enzymes. DNA, found in every plant food for example, requires the enzyme Dnase for digestion. Enzymes also detoxify harmful substances and participate in thousands of never-ending chemical changes in the body. Powdered grass products are carefully processed to preserve as many enzymes as possible. Fresh squeezed wheatgrass juice is a veritable enzyme soup, its cells dancing with metabolic activity. It is so charged with 'chi' electrical energy, you can feel it rushing through your body or raising the hair on the back of your neck. Wheatgrass juice is liquid sunshine transformed into nutritive energy. A veritable brew of water, oxygen, enzymes, protein, phytochemicals, chlorophyll, carotenoids, fatty acids, trace minerals, all rushing to revitalize you.

Nutrition Charts

The following nutritional charts represent comparisons between grass, vegetables and common foods. Where they are available, a variety of different grasses are represented. When you study these charts, keep in mind some of the themes that run through this chapter (*see p. 41–42*). 1) Grass is more than the sum of its parts; 2) Micro-managing our body chemistry—nutrition by the numbers—is not a good idea; 3) Health is a matter of balance and quality, not high milligram counts; 4) Nutrient amounts are in constant flux anyway because of numerous variables.

Lab reports can be confusing. If you can read and understand these data charts, then this book has done its job. First of all, different labs use different units of measure even for the same things. You may find either IU–international units or mcg–micrograms for elements such as beta-carotene and vitamin E. Some use metric system terms like ML–milliliters while other use ounces and teaspoons. Some measure according to serving size while others use the 100g (gram) standard. Converting all the different lab reports to one standard measure is a behemoth job and beyond the ken of the lay consumer. Thankfully, this is something the U.S. Dept. of Agriculture does for us. First of all, they have one standard test for everything. Secondly, they are an independent third party so there is no room for tampering. They also test several lots and average the results. Unfortunately, they haven't yet discovered wheat grass and their vegetable tests don't even include chlorophyll. This book also attempts to average the results of the wheatgrass data in cases where there is more than one report.

No Brands

It is not the goal of this book to give you nutritional analyses of different brands and these charts do not necessarily represent commercially available products. Refer to the nutrition chart on the bottle itself for this information or contact the manufacturer directly.

Analysis of Indoor Grown Fresh Wheatgrass Juice per Ounce					
Moisture	95%	**Vitamin A**	122 IU	**Phosphorus**	21.4mg
Protein	1.9–2.8%	**Vitamin B-12**	.3mcg	**Magnesium**	8mg
Sugars	2–3%	**Vitamin E**	4.3 IU	**Iron**	.66mg
Chlorophyll	4-12mg	**Vitamin C**	1mg	**Calcium**	7.2mg
Biotin	2.86mcg	**Folic Acid**	8.3 mcg	**Potassium**	42mg

Source: Irvine Analytical Laboratory report courtesy Optimum Health Institute.

Comparisons of fresh squeezed juice to dried powder juice are not applicable because fresh juice is 95% water. Additionally, it is too perishable to be perfectly measured. The advantage of the dried grass is that it is stable and measurable. The numbers on the analysis of fresh grass juice do not do it justice. Dr. Ann Wigmore used to say "drink it within minutes of juicing." Greenhouse grass growers who have done lab analyses have been frustrated by the inability to get laboratories to cooperate in this spontaneous fashion. Stories abound of how labs have let it sit refrigerated for a few days or froze it. It is almost a case of the grass telling us we cannot quantify its magic on paper.

Moisture figures are included in the next chart because they have a significant relationship with all other nutrient counts. As you can see, the tray-grown fresh wheatgrass juice is 95% water and about 2% protein. If you were to remove the water, it would be 40% protein.

Chlorophyll and Protein in Green Foods			
Units are % of total content. All grasses except 'fresh' are powders	Moisture	Protein	Chlorophyll
Fresh Wheatgrass Juice	95.0%	1.9-2.8%	0.1%
Whole Leaf Wheat Grass	7.0%	25.0%	0.6%
Kamut Wheat Grass Juice	5-11.5%	19-27%	0.3-2.4%
Alfalfa Juice	0.3%	38.0%	0.7-1.7%
Barley Grass Juice	1-8%	27-46%	1.5%
Freeze Dried WG Juice	8.0%	43.0%	0.1%
Spirulina	3-7%	60.0%	1.0%
Chlorella	5.0%	55-65%	2-3%
Blue Green Algae	2.0%	62.0%	1-2%

Source: Compilation of various reports. 'WG' is tray-grown wheatgrass juice.

That is why the moisture counts are listed. The more moisture, the lower the nutrient numbers. Ranges are given wherever multiple analyses were obtained.

The Seventy Percent Chlorophyll Myth

There is a common misconception from the 1970's that fresh wheatgrass juice is 70% chlorophyll. In fact, it is 95% water and 2% protein. In the remaining 3% lies over 80 other nutrients. Whole leaf dried wheatgrass has 543mg chlorophyll per 100 grams, less than 1%. This nutrient does not comprise a large percentage of any green plant, even though grasses, along with alfalfa and the algaes, are the finest sources of chlorophyll on the planet. This pigment is potent at this dosage. It just doesn't take much of the stuff to perform wonders.

Acknowledgments

The grasses used in these analyses are tray-grown fresh squeezed wheatgrass juice, whole leaf dehydrated grass, wheat and barley grass juice powders, and Kamut juice powder. All data is reported by certified laboratories. The analyses were provided courtesy of: Pines International,

Green Foods Corp. Green Kamut Corp., SweetWheat Inc., VitaRich Foods, The Optimum Health Institute, V.E. Irons, Inc.–Sonne and U.S. Dept. of Agriculture Human Nutrition Research Center. A special thanks is given to these organizations. To find out more about these sources see chapters: *The Companies, Healing Resorts,* and *The Pioneers.*

Vitamin & Mineral Comparison of Grass & Common Foods

per 100 grams		**Grass**	**Sprouts**	**Spinach**	**Broccoli**	**Eggs**	**Chicken**
Protein	g	25%	7.490	2.860	2.980	12.440	17.550
Fat	g	7.980	1.270	.350	.350	9.980	20.330
Calcium	mg	321.000	28.000	99.000	48.000	49.000	10.000
Iron	mg	24.900	2.140	2.710	.880	1.440	1.040
Magnesium	mg	112.000	82.000	79.000	25.000	10.000	20.000
Phosphorus	mg	575.000	200.000	49.000	66.000	177.000	172.000
Potassium	mg	3,225.00	169.000	558.000	325.000	120.000	204.000
Sodium	mg	18.800	16.000	79.000	27.000	280.000	71.000
Zinc	mg	4.870	1.650	.530	.400	1.100	1.190
Copper	mg	0.375	.261	.130	.045	.014	.074
Manganese	mg	2.450	1.858	.897	.229	.026	.019
Selenium	mcg	2.500	n/a	1.000	3.000	30.800	n/a
Vitamin C	mg	214.500	2.600	28.100	93.200	0.000	2.400
Thiamin	mg	0.350	.225	.078	.065	.049	.114
Riboflavin	mg	16.900	.155	.189	.119	.430	.167
Niacin	mg	8.350	3.087	.724	.638	.062	6.262
Pantothenic	mg	0.750	.947	.065	.535	1.125	.920
Vitamin B-6	mg	1.400	.265	.195	.159	.118	.330
Folate	mcg	1,110.00	38.000	194.400	71.000	35.000	6.000
Vit. B-12	mcg	0.800	0.000	0.000	0.000	.800	.320
Vitamin A	IU	513.000	0.000	6715.00	n/a	632.000	178.000
Vit A, RE	mcg	2,520.00	0.000	672.000	154.000	190.000	52.000
Vitamin E	mg	9.100	.050	1.890	1.660	1.050	n/a

Grass used is whole leaf dehydrated wheat grass from Southern Testing & Research Laboratories courtesy Pines International. Sprouts are three day old wheat sprouts. Sprouts and vegetable data from USDA Nutrient Data Laboratory Release 11-1.

Key Nutrients in Different Dried Grasses				
All units in Mg/100g unless indicated	**Whole Leaf Wheatgrass**	**Freeze Dried Wheatgrass Juice**	**Kamut Wheat Grass Juice**	**Barley Grass Juice**
Phosphorus	575	2,700	249	594
Magnesium	112	400	n/a	225
Calcium	321	450	1,110	718
Potassium	3,225	250	4,970	2,762
Beta-Carotene	15,230 IU	4,140 IU	8,940 IU	39,680 IU
Vitamin E	12 IU	2.2 IU	7.24 IU	16.2 IU
Vitamin B-12	2.38mcg	10mcg	n/a	n/a
Vitamin C	215	17	113	132
Iron	25	n/a	n/a	16
Zinc	5	6	4	7

Compilation from various reports. See acknowledgments.

Variables

Variation is the underlying theme when evaluating lab reports. Even within one company and for one product, nutrient counts can vary dramatically from season to season and report to report. Even if the very same batch is brought into two certified laboratories, it brings different results. We tend to think that science is immutable. But testing equipment and environments differ and this natural product, especially when fresh, changes molecule by molecule and minute by minute. With incubation times as long as 20 hours for some tests, it is a race to quantify nutrients before they perish. All we can do is look at averages. Dried powders are much more stable and thus testable than fresh juice. If there ever was a time not to buy according to price, this is it. Go with a reputable brand in which you have faith.

8 Essential Amino Acids of Grass & Common Foods							
per 100 grams		**Grass**	**Sprouts**	**Spinach**	**Broccoli**	**Eggs**	**Chicken**
Threonine	g	1.360	.254	.122	.091	.597	.726
Isoleucine	g	1.450	.287	.147	.109	.679	.877
Phenylalanine	g	1.820	.350	.129	.084	.661	.682
Arginine	g	2.030	.425	.162	.145	.746	1.099
Alanine	g	2.280	.295	.142	.118	.693	1.021
Aspartic acid	g	4.310	.453	.240	.213	1.250	1.565
Glutamic acid	g	4.350	1.871	.343	.375	1.626	2.568
Proline	g	1.570	.674	.112	.114	.496	.846

Wheatgrass used is indoor grown freeze dried 42.8% protein, 8.4% moisture. Analysis by Northeast Labs courtesy of SweetWheat, Inc. Sprouts are 3 day old wheat sprouts. Vegetable data from USDA Human Nutrition Research Center Nutrient Data Lab Release 11-1.

Supplements vs. Whole Food Concentrates

If you want an always dependable milligram amount for a specific nutrient, then buy supplements. The disadvantage of supplements is they are usually isolated nutrients or man-made combinations of them. Although they have many benefits, the advantage of grasses and other superfoods is they are nature-made whole food concentrates, not man-made nutrient selections. The dosages in whole foods may not be as high as your vitamin supplements, but they are significant nonetheless. Whole foods have a qualitative advantage because they are naturally synergistic. This enhances the biological availability of their nutrients, increasing assimilation.

One of the themes of this book is the opposition to nutrition by the numbers. These lab reports show that while there is much nutrition in powdered grass, it is harder to pinpoint the nutrition in fresh juice. Fresh products are simply less stable and yet it should be your nutritional goal to eat more fresh, organic foods. Processing and powdering is not a perfect world, but it is convenient. The trade off for bottled grass is time and effort. You don't have to clean your juicer or wash your vegetables or shop for organic produce or grow anything. That has its value.

While the nutrition information in this chapter is important, the people who have been drinking wheatgrass juice for decades don't need nutrient analyses. They take wheatgrass on faith. For more on why wheatgrass has value beyond science and nutrition, see *Science and Wheatgrass.*

Only by understanding the wisdom of natural foods and their effects on the body shall we attain the mastery of disease and pain.—William Harvey, 1578–1657. English physician who discovered circulation of the blood and the role of the heart.

Research

With each new discovery we learn it's already in the grass.

Nutritional advances are usually developed by examining foods and identifying and isolating factors in those foods. These may be vitamins, amino acids, minerals, enzymes or phytochemicals. Some effects cannot be tied to known nutrients and thus remain unnamed until they are thoroughly verified and characterized. Such is the case when you read about "the grass juice factor." Initial studies are usually carried out on animals like rats, guinea pigs, chicks, rabbits or hamsters. Epidemiological studies map trends on populations and groups. Others test substances in laboratory cultures (in vitro) against other materials including possibly human cells and blood. The ultimate test is a human study where real people take the product. To learn more about the relationship between science and wheatgrass as well as *"How to Study a Study,"* see the chapter: *Science & Wheatgrass.*

Acknowledgments are in order here, especially to the Green Foods Corporation and Yoshihide Hagiwara. As a medical doctor and pharmacist, Hagiwara has matched his considerable business skills with his talents as a scientific researcher. The Hagiwara Institute of Health is the largest privately funded health center in Japan. As consumers, we are fortunate to benefit from his lifelong contribution. Although there is sometimes suspicion and cynicism about whether a corporation is spending such money just to sell more product, that should be evaluated on an individual basis. In this case, Hagiwara is 80 years old and has all the privileges life can offer. Still, he continues to do research.

Barley grass dominates the research in this chapter because of Hagiwara's prodigious work. But barley, wheat, oats, Kamut and rye all belong to the same (triticum) family. Many claim that barley is the superior triticum. However, the Wigmore-based healing clinics and retreat centers achieve their results only with wheat, and most agriculturalists comment that the differences between the grains are minor. True, but when we get down to micro-nutrient levels, the only way to know for sure is to conduct an analysis. You can make your own decisions about which grass is best for you. This book, despite its title, embraces all grasses with equal enthusiasm.

Growth Stimulating Factor & Complete Food

The Relation of the Grass Factor to Guinea Pig Nutrition. By G.O. Kohler, C.A. Elvehjem, and E.B. Hart, Department of Agriculture Chemistry, University of Wisconsin, Madison, Pub November 24, 1937 in the Journal of Nutrition, Vol.15, p. 445. No.5. Also, The Grass Juice Factor, by G. Kohler, S. Randle, and J. Wagner. Journal of Biology and Chemistry. V.128. 1939.

Research into the nutritional content of cereal grasses and curiosity about their secret health promoting factors began in the USA in 1935. A group of scientists from the Department of Agricultural Chemistry at the University of Wisconsin ran a series of experiments in an attempt to learn why milk produced by cows in the winter on winter rations was markedly inferior to milk produced by cows grazing in the spring and summer on fresh pastures. They began adding grass juice to the winter milk and monitored the results on guinea pigs since they, like cows, are herbivorous animals. They concluded that "the growth stimulating factor of grass was distinct from all the known vitamins." They tested three grasses–barley, wheat, and oats–in the dried juice form. All were grown outdoors in the same field at the same time, subject to the same conditions, and all harvested after 30 days. According to Messieurs Kohler, Elvehjem, and Hart:

"The barley grass, which was the most effective, produced a growth rate of 5.3 gm. (weight) per day from the second to the seventh week of the experiment. The wheat grass was only slightly less potent than the barley grass. (...) After 7 weeks on the experiment, pigs(...) were taken off the supplement and fed mineralized milk alone. Growth stopped almost immediately. (Upon resumption) again remarkable growth resulted in the animals receiving the barley and wheat grasses. These results show definitely that the grasses contain a nutritional factor which is essential for maintenance as well as growth of guinea pigs."

The pigs receiving the oat grass did not do as well. However, the wheat and barley juice was so potent that it appeared to be all the animals needed to sustain life. This is surprising because their digestive tracts are equipped to handle large amounts of roughage.

The animals receiving mineralized milk, orange juice and grass juice grew at a good rate and no abnormalities were observed. However, once the grass juice was omitted, the animals died.

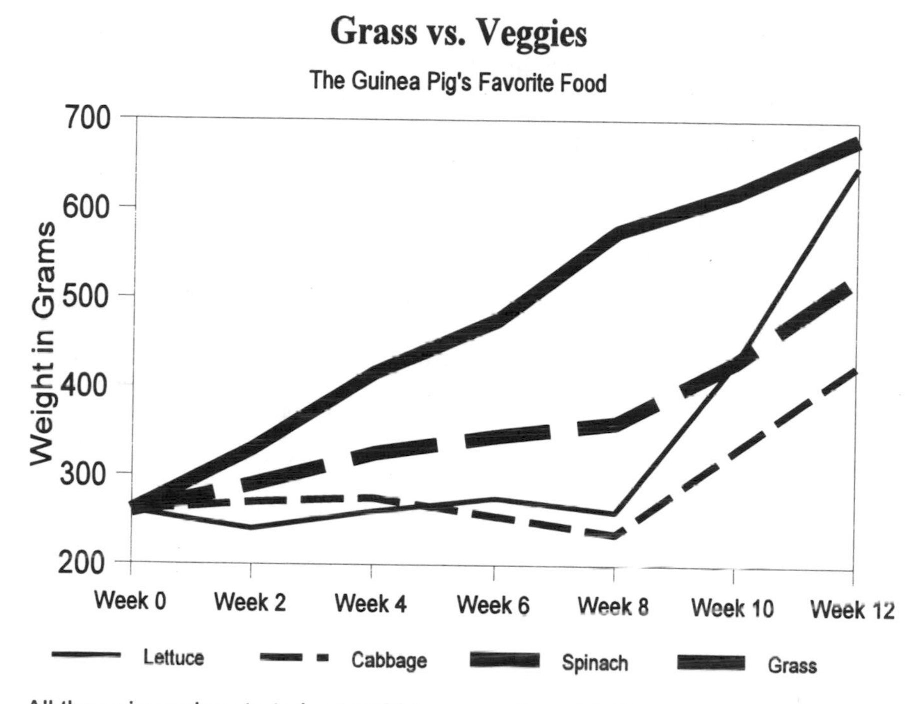

All the guinea pigs started out at 260 grams of weight. One group ate only lettuce, another only cabbage, a third only spinach, and the fourth only grass. On week 8 the cabbage pig was dying and grass was added to all the pigs' diets. All improved dramatically. Spinach was second best. The grass used was only 20% protein.

Early Studies On the "Grass Juice Factor"

1: Growth Stimulating Properties of Grass Juice. By G.O. Kohler, C.A. Elvehjem and E.B Hart. Science, 83:445. 1936. 2: The Relation of the Grass Factor to Guinea Pig Nutrition, by G.O. Kohler, C.A. Elvehjem, C.A. and E.B. Hart, Journal of Nutrition, 15:445. 1938. 3: M. Cannon, and G. Emerson, Journal of Nutrition, 18:155. 1939. 4: Distribution of The Grass Juice Factor in Plant and Animal Materials, by S.B. Randall, H. Sober, G.O. Kohler. Journal of Nutrition, 20:459, 1940. 5: Proceedings Cornell Nutrition Conference, M.L. Scott, p.73, 1951. and Poultry Science, 30:293.

In 1936, Kohler, Elvehjem and Hart reported that fresh grass juice contained an unknown factor necessary for the normal growth of rats. The rats were fed milk produced by cows that grazed on young spring pasture which was shown to be a good source of the water soluble "grass juice factor."[1] In 1938 using the guinea pig as his test animal, George Kohler found that the "grass juice factor" had an important influence on both normal and reproductive growth.[2] His findings were confirmed in 1939 by Cannon and Emerson,[3] and in 1940, the "grass juice factor" was also found in the young plants of white clover, peas, cabbage, turnip tops and spinach. The concentration of the factor was highest in the young grasses and reduced considerably as the grass matured. Other healthy foods such as liver meal and whey had none of the factor and brewers yeast and wheat germ showed only a little.[4] Experiments continued with turkeys and chickens, proving that these animals required the same unknown factor as guinea pigs and rats.[5] Baby chickens fed fresh grass juice produced a "marked growth promoting effect."

Immune Function & Growth Hormone Stimulation

Isolation of a Vitamin E Analog from a Green Barley Leaf Extract That Stimulates Release of Prolactin and Growth Hormone from Rat Anterior Pituitary Cells in Vitro. By M. Badamchian, B. Spangelo, Y. Bao, Y. Hagiwara, H. Hagiwara, H. Ueyama, and A. Goldstein. Journal Nutrition and Biochemistry. Vol. 5: 145-150. 1994. Also, a-tocopherol Succinate but Not a-tocopherol or Other Vit. E Analogs, Stimulates Prolactin Release from Rat Anterior Pituitary Cells in Vitro. By same authors (except Bao). Journal Nutrition and Biochemistry. Vol. 6: 340-344. 1995.

White blood cells protect our bodies from infection and disease and are the front line defense of our immune system. The pituitary gland secretes hormones which control other endocrine glands and influences growth, metabolism and maturation. Barley grass leaf extract (BLE) was added to cultures of human white blood cells and to cultures of anterior (front) pituitary cells. The grass extract stimulated immune functions and the release of growth hormone and prolactin. These hormones are related to general health, reproductive health and other important physiological functions. This study successfully isolated and identified the molecule in barley grass responsible for this enhancement. It is a water soluble form of vitamin E called a-tocopherol succinate that is abundant in BLE. According to Dr. Allan Goldstein co-author of the study, "we now know that vitamin E plays an important metabolic role in maintaining the integrity of membranes and may reduce the risk of heart disease and lower the incidence of several types of cancer in humans, including breast and colon."

Other commercially available forms of vitamin E (alpha-tocopherol, a-tocopherol acetate and a-tocopherol nicotinate) were tested, but only a-tocopherol succinate stimulated the production of prolactin. This micronutrient suppressed human tumor cell growth in cultures where other tocopherols were ineffective. "This suggests a role of alpha-tocopherol succinate as an anti-tumor proliferative agent and as a modifier of human leukemia cell differentiation."

Immune System

Immune System Components Altered by a Food Supplement. Presented by E. Wagner and R. Mocharla, at the 1991 Annual Meeting of the American Association for the Advancement of Science, Feb. 16, 1991. Washington, D.C.

This double blind human study involved 32 first and second year medical students 16 of whom received 6 grams of barley grass leaf extract (commercially sold as Green Magma) for 71 days. The other 16 received placebos. Everyone had a complete blood analysis before and after in which various components of the immune system were measured. The students who received the barley grass juice supplement exhibited a statistically significant increase in their leukocytes and a decrease in their lymphocytes. Their overall immunity was stronger than the placebo group. The authors commented that although no definite conclusions can be drawn, "one could hypothesize that the statistically significant increases in the percent of neutrophils and circulating complement levels, both components of nonspecific immune defense in the nutritionally supplemented group, could reflect a more efficient first line of defense."

Wheat Sprouts Increase Antioxidant Levels

Effects of Whole Live Foods on (SOD) Deficiency in 10 Adult Humans. Conducted by Dr. Peter Rothschild M.D. Ph.D. et al. Testing by Smith-Kline Bio-Science, Honolulu, HI. Antioxidant enzymes supplied by Biotec Food, HI. Courtesy of AgriGenic Food Corp. Huntington Beach, CA.

Ten senior citizens had their blood levels of superoxide dismutase (SOD) tested before and after taking a hydroponically grown wheat sprout supplement. The supplements were manufactured by AgriGenic Food Corp. *(see resources)* from the sprouts, a pre–grass stage of growth, using a low temperature process to preserve enzyme activity.

Young grasses are excellent sources of superoxide dismutase (SOD), a powerful antioxidant and anti-aging enzyme. This human study used 70 year old (average age) seniors who are naturally slower to respond than a younger population. Nevertheless, the serum levels of the seniors increased an average of 230% overall in the group and as much as 730% in one individual. Seven of the 10 participants more than doubled their blood levels of SOD. This proves that wheat sprouts are an excellent source of SOD and that potency can be maintained in a supplement. The high levels of assimilation are likely due to the fact that the supplement is a whole foods concentrate with all of its synergistic co-nutritional factors.

A Powerful Antioxidant

Inhibitory Effect of 2"-0-Glycosyl Isovitexin and a-Tocopherol on Genotoxic Glyoxal Formation in a Lipid Peroxidation System. By T. Nishyama, Y. Hagiwara and T. Shibamoto. Dept. of Environmental Toxicology, University of California, Davis. Published by Food Chemical Toxicity, Vol. 32, No. 11, pp. 1047–1051, 1994. See also the original study: A Novel Antioxidant Isolated from Young Green Barley Leaves, Agricultural and Food Chemistry. Vo. 40, pp. 1135-1138. July, 1992.

This study tested fractions of young barley grass juice powder on highly toxic oxidizing chemicals to test the grass' ability as an antioxidant. Oxidation damage is associated with aging, cancer, HIV and other immunodeficiency diseases. Oxidizing agents reduce healthy compounds into toxic ones. When fats and oils (lipids) are oxidized, they degrade into rancid fats that react with amino acids and proteins and become hazardous to human health. This research examines glyoxal, a potent mutagen common in cigarette smoke. Living cells depend on enzymes and dietary antioxidants to protect themselves. Some famous antioxidants are vitamin C, beta-carotene and vitamin E (alpha-tocopherol), but there are many others in plants.

In this experiment, an antioxidant named 2"-0-Glycosyl Isovitexin (an isoflavonoid abbreviated 2"-*0*-GIV for short) was isolated from young green barley leaves. Its capacity to prevent the formation of glyoxal from degrading fats was measured. A kind of vitamin E, a-Tocopherol was also tested as a comparison. The amount of glyoxal formed from three different fats (fatty acids) in the presence of each antioxidant was measured. This was all done in vitro, in a test tube, not in an animal and the results were replicated at least two times.

2"-*0*-GIV inhibited glyoxal formation by nearly 70%. "Alpha-tocopherol exhibited a greater inhibitory effect at the lower levels. On the other hand, 2"-0-GIV showed a higher effect than a-tocopherol at the higher levels. 2"-0-GIV is more effective than a-tocopherol towards fatty esters with higher numbers of double bonds. The more bonds possessed by a lipid, the more oxidation products it produces. A maximum inhibition of 82% was obtained by 2"-0-GIV." The antioxidant activity of isoflavonoids such as 2"-0-GIV was hypothesized because of their ability to chelate metal ions and scavenge free radicals. "In addition, flavonoid compounds reportedly affect many biological processes. They exhibit anti-hepatotoxicity, anti-inflammatory effects, anti-allergy effects and antiviral activity."

Anti-Ulcer

Anti-ulcer Activity of Fractions from Juice of Young Barley Leaves. By H. Ohtake, H. Yuasa, C. Komura, T. Miyauchi, Y. Hagiwara and K. Kubota. Issued from Pharmaceutical Society of Japan.

Until now, no evidence has been reported that pharmacologically demonstrates that grass juice can prevent peptic ulcers. This experiment created stomach ulcers in rats and then cured them with barley grass juice in 3–4 days. Rats were induced with different types of ulcers, treated with grass juice fractions, then cut open and the condition of the ulcers inspected. Seven different fractions of spray dried, green barley grass powder were tested along with a control.

Green barley juice revealed significant anti-ulcer activity in the various ulcers and one fraction "P4D1 significantly promoted the healing of the gastric ulcer induced by an injection of dilute ascetic acid into the gastric wall of the rats." On stress induced ulcers, all fractions of the barley grass exerted "significant anti-ulcer effects." The water soluble fraction GM-L exhibited the highest activity on stressed induced ulcers. Three fractions proved significant anti-ulcer agents against aspirin induced ulcers. None of the grass factors influenced aspirin absorption, gastric secretion of acid and pepsin. So their anti-ulcer action is not a function of suppressed gastric secretion.

The fractions were composed of water soluble substances of chlorophyll, denatured protein, polysaccharides, amino acids, high molecular substances and insoluble substances. "It is unlikely that green barley juice exerts its anti-ulcer action via affecting the attacking factors of the ulcer, but suggested that the effects on the movement and blood flow in the stomach and the defense ability of the stomach mucosa are associated with its anti-ulcer activity."

Antidote for Food Additives and Insecticides

Effect on the Several Food Additives, Agricultural Chemicals and Carcinogen. By Y. Hagiwara, M.D. Presented to the 98th Annual Assembly of Pharmaceutical Society of Japan, April 5, 1978.

This paper presented the results of Dr. Hagiwara's tests on the ability of grass juice to inactivate mutagenic substances found in agricultural chemicals, fertilizers and food additives.

Barley juice extract successfully decomposed the agricultural insecticide Malathion, dropping its concentration down from 100 parts per million (ppm) to 19ppm in two hours. The food additive sorbic acid decomposed from 100 ppm down to 9ppm after 12 hours. The synthetic antioxidant and food preservative BHT was decreased from 400ppm to 40ppm after 5 hours of incubation with barley grass extract.

Stamina and Endurance

Studies on the Effects of Green Barley Juice on the Endurance and Motor Activity in Mice. By K. Kubota and N. Sunagane, Faculty of Pharmaceutical Sciences, Science Univ. Of Tokyo, Japan. Presented Before the 104th Annual Congress of Pharmaceutical Society of Japan, Sendai, 1984.

This study tested the motor activity and endurance rates of mice who had spray dried barley grass juice added to their standard diets. 32 mice were tested, 16 with 2% grass added to their food and 16 without any as a control. The number of revolutions on their wheel cages was recorded over a 2 hour test period each day for 15 days. The grass fed mice revolved the wheel cage more than the control mice in numbers that were "statistically significant." (Right chart)

On the endurance test, (left chart) 16 mice were prodded to race on an uphill sloped treadmill belt after 15 days of 2% and 4% barley grass added to their diet. "The momentum (body weight multiplied by duration) in the mice fed on 4% barley was significantly larger than that in the control group." "The average body weight of mice fed on the barley grass food was larger than the mice fed on the standard food."

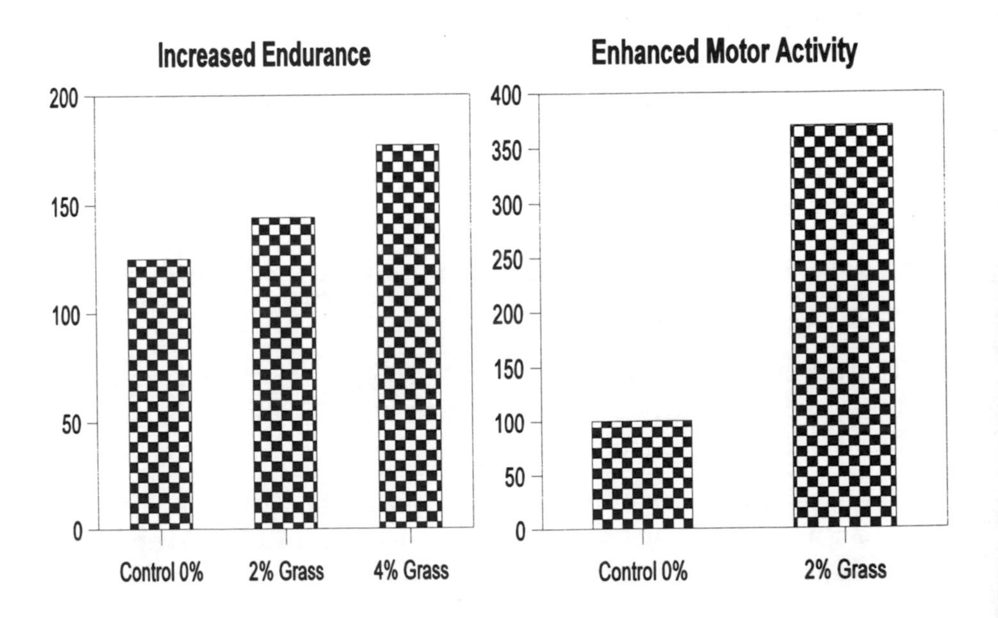

Dr. Ann Wigmore

"I separated six day old chicks into two groups of three chicks each. Each group was fed the best accepted type of chick food. But in one cage I mixed chopped up, freshly gathered wheatgrass with the food and placed a sprig of wheatgrass in the drinking water. At the end of a few weeks, all the chicks were healthy, but those receiving the wheatgrass had grown twice as large as the others. They were more alert and had feathered out better. Groups of rabbits and kittens, fed in similar fashion showed the same results in size, weight, and mentality."

Skin Diseases

Therapeutic Experiment of Young Green Barley Juice for the Treatment of Skin Diseases in the Main. By Tatsuo Muto, Muto Dermatologic Hospital. New Drugs and Clinical Application. Vol. 26, No. 5 May 10, 1977. Courtesy of Green Foods Corporation.

Treatment of skin diseases should include not only symptomatic therapy but also systemic therapy. In this long term clinical trial, 38 skin disease patients were selected for their resistance to conventional drugs. They had everything from acne to eczema to atopic dermatitis. A typical dose was 4 grams of green barley grass juice powder dissolved in water three times daily, either in between meals or 30 minutes before them. In addition to observations, blood, urea and liver function were tested. Considerable effects were noted in almost all cases in 1 to 6 months. "What was certain is that the administration promoted blood circulation, appetite, and bowel movement and the patients complexions improved." One 43 year old male with gout and viscous eczema on his back "had pain in his right leg and his uric acid was a high 8.6. After two months, it dropped to 6.5 and his eczema was completely cured." It even had an effect on leukoplakia, white spots in the mouth. "The ratio of effectiveness was 81.6% in all cases." There were no side effects. The author concludes that overall, barley grass powder gives "excellent results cosmetically and in the prevention of aging."

The strong restorative power of green barley powder which is effective for preventing cancer, has the added advantage of making our skin beautiful. Dry, rough skin is associated with aging.

—Yoshihide Hagiwara

Longevity

Wild-growing Edible Grasses in the Nourishment of Long Living Inhabitants of the Dagestan Republic. By Kh.I Mustafaev. Published in the Russian language periodical Voprosy Pitanieiia. Moskva: Biomedgiz. Vol.2, no.5. pp.27-31.Sept/Oct. 1993.

This study examined the diets of the inhabitants of the Dagestan republic who are known for life spans of 120–130 years. The oldest recorded Dagestani lived to 146. The 2 million residents of this province live in the Caucasus mountains of southern Russia and east of Georgia on the coast of the Caspian Sea. Wheat is this country's chief crop and their diet consists of many wild grasses and weeds such as chickweed, shepherd's purse, rose hips, camomile, lambs quarters, thistle, thyme, sorrel, yellow dock, vetch, daisy, clover, wild marjoram, oregano, amaranth, mustard, garlic, and the grasses of wheat, barley, and oats. They use the young leaves to make a raw salad and boil the older fibrous ones for soups and stews. Seeds are crushed and brewed into tea or ground into meal and used in breads and pancakes. They also make pickles and sauerkraut and, yes, yoghurt.

Researchers from the Caspian medical college examined 154 alpine residents living at altitudes of 6,400 ft above sea level and 24 living on the flat lands. Their families were also observed. The age of the test group was between 85 and 116 years old. Researchers lived with their subjects for 10–12 days, questioning them, weighing them and examining their diets and eating habits.

The "long livers" wake between 5–6am and drink nothing but tea made from weeds and grasses until they take breakfast at 9–10 am, never before. They drink tea both before and after the meals for increasing appetite and improving digestion. The 90 year olds had the most raw greens in their diet of any age group but even the children ate greens. Researchers found the wild growing edible plants to be rich in B vitamins, citric acid rose hips, vitamin C and A and pectins. The plants contained other nutritional substances that were excellent natural stimulators of metabolism and digestion.

The flat-landers were not as healthy as the alpines but neither group had any signs of heart disease or hypertension. Researchers concluded that "many factors influence health including lifestyle, genes, heritage and work...(but) the location and altitude along with their unusual diet rich in young wild grasses, produced their longevity."

Cellular Rejuvenation, DNA Repair, Anti-Aging

Preliminary Report on How Juice of Young Green Barley Plants Can Normalize and Rejuvenate Cells and Tissues, Repair Damaged DNA, Restore Cellular Activity and Prevent Aging of Tissues. By Y. Hagiwara, Y. Hotta, K. Kubota. Japan Pharmaceutical Development and Biology Dept. Univ. of CA, San Diego. Reported to Annual Japan Pharmacy Science Assoc. Meeting.

Damage to our genes (DNA) can be caused by many factors including everything from agricultural chemicals to medical drugs, radiation, x-rays, lack of enzymes and stress. According to Dr. K. Kubota of the Tokyo Pharmacy Science University: "A special fraction called 'P4D1' from green barley juice produced a remarkable stimulation to the repair of cellular DNA." Meiotic cells (the reproductive cells) are equipped with DNA repairing enzymes and binding proteins that can repair any damage to the DNA. Kubota isolated the meiotic cells and damaged them with x-ray irradiation and quinoline, both of which are carcinogenic. "When these cells were incubated under normal conditions, repair to the DNA was slow and some cells died. However, when P4D1 was added into the culture, the repair of DNA was promoted significantly both in time and quantity." No side effects were observed.

Activity of the meiotic cells decrease with age. However, "it is exciting to discover that this reduction in meiotic activity with age can be remarkably restored by the fraction P4D1. No stimulation of DNA repair or promotion of meiotic activity has been reported earlier from the use of any natural or synthetic product."

Inhibition of Carcinogens

Inhibition of In Vitro Metabolic Activation of Carcinogens by Wheat Sprout Extracts. By Chiu-Nan Lai, B. Dabney, C. Shaw, Dept. of Biology, Univ. of Texas System Cancer Center. M.D. Anderson Hospital and Tumor Institute, Houston, TX. Nutrition and Cancer. Vol.1, no. 1. P.27-30. Fall, 1978.

This experiment took commercially available wheat berries and grew them hydroponically for 7–14 days into five inch tall wheat grass. Wheat-grass juice was extracted from the roots and leaves and tested against known carcinogens requiring metabolic activation. Extracts from carrots and parsley also exhibited inhibitory activities but not as potent as those of wheat. Unsprouted wheat berries soaked overnight did not demonstrate any inhibitory activities.

"These results are of interest for two reasons: first, the inhibition of activation of potent carcinogens is quite strong at a reasonably low level of extract and second, the wheat sprout extract is nontoxic even at high levels while most known inhibitors are toxic at medium to high levels." *(See also, "A Powerful Antioxidant.")*

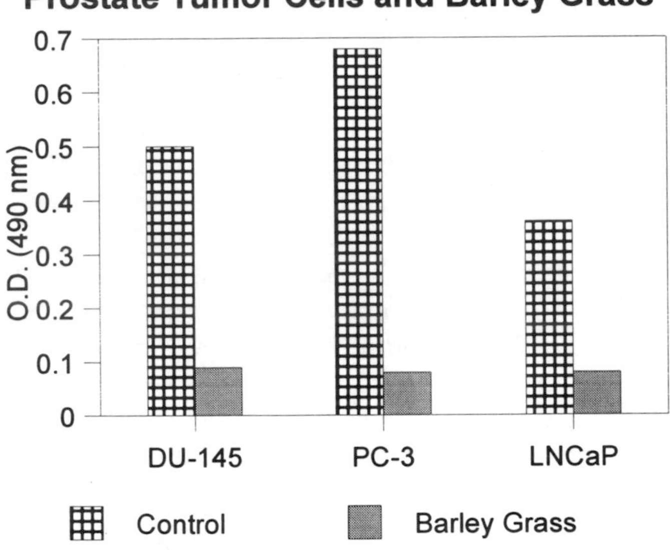

Anti-Cancer

Biochemical Characterization of Novel Molecule(s) in Barley Leaf Extract That Inhibits Growth of Human Prostate Cancer Cells. Preliminary report. By Dr. Allan L. Goldstein, Ph.D, Professor & Chairman, Dept. of Biochemistry & Molecular Biology. George Washington Univ. Medical ctr.

This chart represents a preliminary report on the action of an unnamed molecule(s) isolated from barley grass leaf extract (BLE) that has powerful anti-cancer properties. When added to a culture of human prostate tumor cells, it inhibited their proliferation. BLE competed against three human prostate cells: PC-3 a prostate adenoma, DU-145 a prostate carcinoma with metastasis to the brain and LNCaP a metastatic prostate adenocarcinoma. The tall bars on the chart are the three cancer cells and the short bars show their levels after the barley grass extract factor was introduced. Dr. Goldstein says: "To date, we have isolated and partially purified a molecule(s) with potent anti-cancer activity from barley grass leaf extract. BLE dramatically inhibits the growth of human prostatic cancer cells grown in tissue culture. The complete purification and characterization of this molecule is a top priority in our lab as it may provide a new nutritional approach to the treatment of prostate cancer that could compliment and perhaps improve standard therapies." *See also: Immune function and alpha-tocopherol succinate.*

Anti-Inflammatory

Isolation of Potent Anti-Inflammatory Protein from Barley Leaves
By K. Kubota, Y. Matsuoka, H. Seki, Faculty of Pharmaceutical Sciences, Science Univ. of Tokyo, Japan. Japanese Journal of Inflammation, Vol. 3, no. 4, 1983.

Superoxide dismutase (SOD) is well known for its potent anti-inflammatory action. Since barley grass is abundant in SOD, this study examined the juice of young barley grass and discovered that the anti-inflammatory effect was assisted by other proteins. These protein fractions were isolated and tested. Named P4D1 and D1G1, they are glyco-proteins that are both heat stable and highly soluble in water. In this experiment, male Wistar rats were induced with edema and then treated with the proteins orally, subcutaneously and by injection. No toxic signs were produced in the rats even at very high doses. Although SOD's effectiveness was significantly reduced with heat, "D1G1 was hardly reduced after heating to 100°C for 20 minutes."

"All of D1G1, P4D1 and SOD isolated from green barley juice revealed extremely potent anti-inflammatory activity, especially when they were injected intravenously. They significantly suppressed the carrageenan-induced edema in rats at very low doses..." Plus, both are chemically different from SOD and aspirin, also famous for its anti-inflammatory effects. "P4D1 and D1G1 seem to be much better anti-inflammatory agents than aspirin as far as they were concerned with intravenous administration."

Barley Grass Eliminates Symptoms of Pancreatitis

Therapeutic Effect of Water Soluble Form of Chlorophyll-a and the Related Substance the Young Barley Green Juice in the Treatment of Patients with Chronic Pancreatitis by Osamu Yokono, M.d. First Dept. Of Medicine, Faculty of Medicine, Univ. Of Tokyo. Courtesy Green Foods Corp.

Pancreatitis is a debilitating disease wherein the pancreas becomes inflamed, causing persistent abdominal pain, nausea, vomiting and high fever. The drug Trasylol is often recommended with mollifying results in the majority of cases but has side effects. While chlorophyll has been proven effective in treating pancreatitis in laboratory experiments and actual clinical trials, it needs to be administered intravenously which is neither convenient nor practical. Getting chlorophyll via foods is difficult because "less than 5% of ingested chlorophyll can be absorbed." The aim of this study was to discover the effect of ingested chlorophyll on these chronic symptoms. The food source was the freeze-dried extract of young green barley grass juice because it contains 1.5 grams of chlorophyll per

100 grams. Twenty-four chronic pancreatitis patients participated in this long term double-blind test.

"From the observations, we had the impression that the curative effect of the young green barley juice was obtained especially in the relief of abdominal pain. As for the so called side-effects of the drug, nothing unfavorable was observed clinically in the results before and after the administration of the drug, in urine analysis, hematologic surveys, liver function tests, kidney function tests and blood coagulation functions." But because barley grass juice contains "numerous known and unknown bio-active substances," these positive results may be skewed by substances other than chlorophyll. "Irrespective of the presence of uncertainty about the identification of the active substances contained in the young green barley juice, our results presented here suggest that the oral administration of the young green barley juice gives some definite favorable effects on the pathological phenomena related to chronic pancreatitis."

Prevention and Cure for Atherosclerosis

Inhibition of Malonaldehyde Formation from Lipids by an Isoflavonoid Isolated from Young Green Barley Leaves. By Y. Hagiwara, T. Shibamoto, T. Nishiyama, H. Hagiwara. Journal of American Oil Chemists Society, Vol. 70, no 8. Aug. 1993. Studies on the Constituents of Green Juice from Young Barley Leaves Effect on Dietary Induced Hypercholesterolemia in Rats. By Y. Hagiwara, K. Kubota, S. Nonaka, H. Ohtake, Y. Sawada. Journal of the Pharmaceutical Society of Japan, Vol. 105, No. 11. 1985. Inhibition of Malonaldehyde Formation by Antioxidants from 3 Polyunsaturated Fatty Acits. By J. Ogata, Y. Hagiwara, H. Hagiwara, T. Shibamoto. JAOCS. Vol. 73, no. 5. 1996

The peroxidation of skin lipids from exposure to UV light has been associated with skin aging and skin cancer. These highly reactive peroxides promote cancer, mutation and aging. The oxidation of fats (lipids) also plays an important role in the development of atherosclerosis, the buildup of cholesterol plaque on arterial walls. These studies show that antioxidants can counteract the oxidation of fatty acids and demonstrate results that are "not only preventative but curative."

In this study, freeze-dried young barley grass juice was extracted and broken down into fractions. The powerful antioxidant 2"-0-Glycosyl Isovitexin or 2"-0-GIV for short, was isolated from the grass juice. Its antioxidant activity was measured using gas chromatography analysis of malonaldehyde (MA), a relative of formaldehyde formed during oxidation of fats. 2"-0-GIV was compared with a form of vitamin E called alpha-tocopherol and BHT, a synthetic antioxidant used to preserve fats and oils in foods. Other comparisons included the high cholesterol drug Probutol. Several different kinds of fats were oxidized in test tubes of human plasma and replicated at least twice.

2”-0-GIV performed competitively with vitamin C, vitamin E (alpha-tocopherol) and BHT across the various fat oxidation experiments. While a-tocopherol scored higher in some tests, it broke down under UV-irradiation while 2”-0-GIV did not. “The antioxidant isolated from young barley leaves is more effective than the other two antioxidants (a-tocopherol and BHT) upon UV-irradiation.” When vitamin C was added to 2”-0-GIV, it became more potent. The drug Probutol achieved slightly better results but by so little that it was declared to be just as effective. Just 2 units of 2”-0-GIV inhibited MA by almost 100% while BHT required 12 units and still failed to match 2”-0-GIV. “The isoflavonoid (2”-0-GIV) demonstrated significant anti-oxidative activity toward lipid peroxidation and furthermore can be obtained in large quantities from a natural source at low cost.”

In the 1985 study published in the Journal of the Pharmaceutical Society of Japan, rats were fed high cholesterol diets and then fractions from barley grass juice extract. Their cholesterol was measured before, during and after. The grass juice fractions significantly lowered their serum cholesterol levels. “The present work suggests that green barley juice may be very useful for preventing humans from vascular diseases associated with hypercholesterolemia.”

Grass Juice or Whole Leaf—Both are Good

1: Proceedings Cornell Nutrition Conference, M.L. Scott, p.73, 1951. and Poultry Science, 30:293. 2: Unidentified Factors in Alfalfa and Grasses, by S.J. Slinger, Feedstuffs, p.8a. Jan. 21, 1956. 3: Anderson, G.W., Slinger, S.J. and Pepper, W.F. Unpublished results, Ontario Agricultural College. Guelph, Canada 1954. Reported in Feedstuffs, Jan 21, 1956. 4: Hansen, R.G., Scott, H.M., Larson, B.L., Nelson, T.S. and Krichevsky, P.J. Journal of Nutrition, 49:453. 1953.

M.L. Scott's research determined that the grass factor was rich in the juices of grass and alfalfa but was largely, though not entirely, destroyed by dehydration.[1] Dr. S.J. Slinger, commenting on the issue of whole leaf vs. fresh juice powder noted that “while the activity is partially destroyed by dehydration, there appears to be a significant amount of the factors present in certain dehydrated alfalfa and cereal grass meals. [But] juice preparations are more consistently potent sources of the activity than dehydrated products."[2] One study made a direct comparison of dehydrated whole cereal grass, sun-cured alfalfa and grass juice concentrate. They found that the grass juice, from a mixture of various field grasses and alfalfa, was superior to the whole dried cereal grass but not significantly so. Turkeys and other poultry grew equally well, indicating that the factor was present in the whole dried grass.[3] Dehydrated or sun-cured alfalfa gave a response at a dosage of 10%–20% comparable to that of the grass juice at a dosage of 5%.[4]

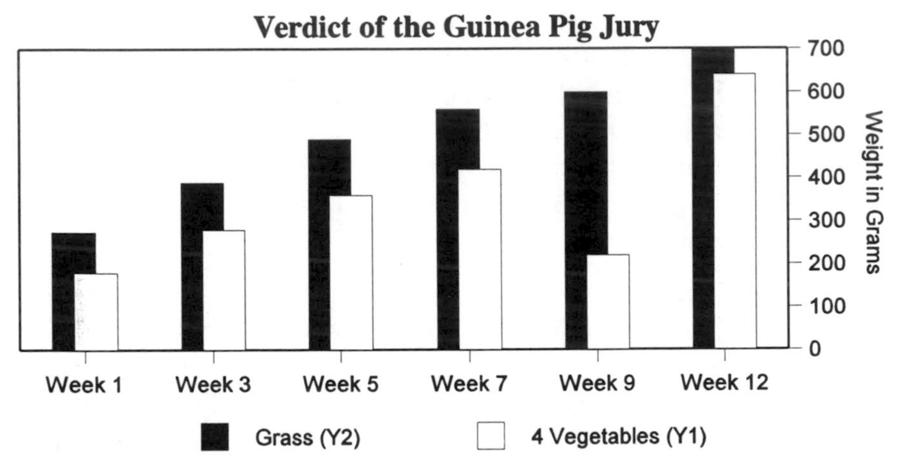

The Relation of the Grass Factor to Guinea Pig Nutrition. By G.O. Kohler, C.A. Elvehjem, E.B. Hart, Dept of Agriculture Chemistry, Univ. of Wisconsin. J. of Nutrition, V.15, No.5. 11/24/1937

Notice that even when four of our best vegetables (spinach, carrot, cabbage and lettuce) were fed the guinea pigs, the growth was erratic and the pigs started to fail rapidly after the seventh week. To save their lives, dehydrated cereal grass was added to their diets (11th week) and in less than a week, the trend was reversed and a consistent and rapid growth took place. The experiments prove conclusively that while other foods are good, the grasses alone are the complete food and contain all the elements needed to support life. —*V.E. Irons (see The Pioneers)*

> *Animals don't know anything about vitamins. They determine the nutritive value with their instinct, palate and olfactory faculties acting for them in place of judgement.* —George Sinclair, 1869.

Healing with Grass

The Great Power Beneath our Feet

An analysis of blood samples drawn from more than two hundred Hippocrates guests before and after the two week program gave scientific support to our observations. Performed at the Arthur Testing laboratory, the study showed that within two weeks of following the Hippocrates live food diet and drinking wheatgrass juice, the blood is detoxified and the immune system strengthened. These changes lead to more energy and an improved ability to combat and reverse illness.

—Dr. Ann Wigmore[1]

Healing is a Journey

Maybe you think that just by taking wheatgrass it will heal you. Well, a car will get you from Florida to Alaska, but it's a long trip and your car better be in good shape. Wheatgrass is detergent for the engine, oil for the motor, lubricant for the axles, grease for the gears and coolant for the radiator. You can't manage a trip like this without it. On the other hand when you arrive, although you may praise your lubricant, the real accolades go to the driver.

However wonderful wheatgrass is, it is only the fuel energizer in your engine. Many who speak about wheatgrass exclaim its virtues and how it saved them. Wheatgrass gets a lot of credit. But it is neither a panacea nor a magic potion and some leaders in the wheatgrass movement will tell you it does not cure anything. People heal, wheatgrass helps. You will restore balanced health as a result of the changes you make. Wheatgrass may get you started on this new path, but you will work very hard at applying a multi-faceted total health restoration program. It can take over your entire life. Sick people are often too stricken to function normally in society. They stop working, have long hospitals stays, are bedridden—they have to drop out to die or drop out to strive to live. Even if you take 8 ounces of wheatgrass juice everyday and build a health program around it, it is only the grease for your axles. To allege that wheatgrass cures is to equate it with a drug. That turns wheatgrass into a quick-fix remedy which is the antithesis of what the health concepts behind it are all about.

The Causes of Disease

Health is a balancing act. Around 1980, world famous tightrope walker Phillipe Petite walked a rope from one tower of New York's World Trade Center to the other 1,400 feet in the air. He took his life in his hands for real. There was no parachute. Were he to make a false step, he would die. Everyday we take steps in our life that either keep us in perfect balance or throw us on a downward spiral. The fall into disease is not a single bad step, but a series of bad choices. Call them cigarettes, drugs, alcohol, junk food, poor hygiene, stress, work environment, air pollution, water pollution, the result is cell pollution. The signs of disease are low energy, fatigue, poor digestion, gain or loss of weight, unclear thinking, allergies, aches and pains and ultimately the major disorders—cancer, high blood pressure, heart disease, arthritis, emphysema, etc. Cells start to lose their life force. They misfire. Metabolism malfunctions. Organs weaken, digestion and elimination are disrupted. Toxins settle into the dead zones—the weak spots. Now the cells start to darken, choking on bacteria, yeast, fungus and toxic acids. They can't get enough oxygen. Acidification, infestation and destruction—the cycle of imbalance. We have fallen.

Profile of A Cancer Patient

In a study of new cancer patients at the Cancer Treatment Center of Tulsa, Oklahoma, most patients smoked or had smoked, did not like to eat green vegetables, had below normal body temperatures, rarely exercised to a sweat, ate large proportions of refined flour products (bread, crackers, cookies, rolls) had a stressful event in their recent history, and last but not least, drank soda pop, colas, coffee, bottled juices, and very little water.

Rebuilding Health

The journey to rebuilding your health is akin to rebuilding your house. First of all, it needs a thorough cleaning. Forget about the mop. Pull out the heavy duty vacuum, rent a floor sander, pull down the old wallpaper, rip out the broken cabinets, reset the squeeky door, spackle the hole, fix the windows. Although, you may need assistance from professionals—the painter, carpenter, electrician—you are the contractor. It's your house; you direct the nature of repairs; you choose the pace and you have to live in it during the renovation. Hopefully, you have not let things go so far downhill that you find structural damage. If you've ignored the signs and have termite infestation for example, you may have to surgically replace a beam. If too many beams have been damaged, then the house may be beyond repair. There is a critical mass in any system beyond which it cannot recover. Although such deterioration does not occur overnight, it did happen right under your nose. You contributed to it by ignoring the signs and symptoms or taking wrong advice from the professionals. Or maybe you've just lived like a slob and dirtied your own house and now look at the mess you're in. It's time to change your ways—renovate, rebuild, renew.

Step I –Cleanse

The path to health starts with a major cleaning. If there is infestation, you have to control it. If there is damage, you have to reconstruct. First stop the poisoning of the blood and tissues from acidifying diet, bacteria, yeast, giardia and parasites. Reverse the trend of acidification and alkalinize the blood with wheatgrass juice and an enzyme rich living foods diet. Take nothing from a box or can. Eat from the garden. Eliminate sugar except when it comes inside a fruit. Throw out the espresso machine and the microwave and put a juicer and blender in their place.

Clear the blockages, empty the garbage and reinvigorate the elimination system with colon hydrotherapy. Jump start the liver with a massage and a wheatgrass implant. You can change. Your body is a milky-way of trillions of cells which combine to make up tissues and glands. But every minute we breath in new atoms and exhale the old. In six weeks, every atom in your liver is replaced. Renewal is a constant. What foods will you renew with? Stop treating your stomach like a compost and start treating it like a garden. Set in motion an upward cycle of rejuvenation that leads to balanced health. The signposts are: trimmer weight, clearer mind, brighter eyes, better concentration, more energy and vitality.

He who is immune is immortal.
—Lao-Tzu, ancient Chinese philosopher, founder of Taoism

Many Diseases, One Cure

Although there are many diseases, all cures start with detoxification. Grasses with their living chlorophyll act in our bodies like detergents purging the liver, scrubbing the intestinal tract and oxygenating blood. Grass chelates and removes heavy metals from our bodies. While hemoglobin has a strong iron bond, chlorophyll has a weak magnesium bond *(see p. 50)*. Thus, wheatgrass readily releases its magnesium, creating a cellular vortex—a cyclone sucking in heavy metals. Tests on fecal matter, done before and after taking grass juice, show higher heavy metal counts after wheatgrass.[2]

Fasting gives your body a chance to re-balance its own chemistry, eliminating all the drugs and supplements that force it on detours and into different directions. Fasting is the great eraser, reducing all outside influences and creating a fresh start. Fasts of 3–50 days have achieved many miracles. Juice fasting with wheatgrass and raw vegetable juices introduces oxygen and nutrients into the bloodstream while enhancing elimination.[3] From here, the reintroduction of 100% fresh foods gives you the potential for normalization and rejuvenation.

A fully functioning elimination system is fundamental to any healing program. Colon hydrotherapy is a must. True, this is a much shunned subject that is rarely accepted for conversation. After all, it is the cess pool of the body. But herein lurks the origin of much disease. The colon is the exit door for putrefactive poisons. If the doorway is jammed, these poisons throw an extra burden back on the liver and start a cycle of autointoxication. Judging by the sales of laxatives, we Americans are a pretty constipated lot. One of the first noticeable symptoms of grass is its laxative effect.

The sweetness of tray-grown grass is actually part of its power. This grass is so high in sugars that it can cause nausea if not taken on an empty stomach. But the sugar helps deliver grass' crude chlorophyll quickly into the bloodstream. Sugars in the intestinal tract crystallize, drawing out embedded toxins from the tissues. The powdered grasses cannot claim this feature because the drying changes the way the sugar crystallizes and they have fewer simple sugars anyway.

The liver is the largest internal organ of the body and the most important organ of detoxification. It detoxifies drugs and removes toxins and waste products from the blood. A functioning liver is required for any health comeback. But too many chemicals from unnatural foods, drugs, poisons, bacteria and parasites can scar the liver (cirrhosis) or infect it (hepatitis) or otherwise overload it into dysfunction. Liver and kidney damage is the first step on a vicious downward spiral of degeneration. Wheatgrass has a purging effect on the liver and implants are the best way to cleanse and rejuvenate this vital organ.

> *The most striking thing was the appearance of their livers which were a dark mahogany color and the surface of the livers glistened like a mirror. The alfalfa fed hens had light tan colored livers. These liver changes caused by good grass are too obvious not to have some connection with the prevention of degenerative disease.*—Dr. Charles F. Schnabel.[4]

Step II Nourish

Once the house is clean, it's time to rebuild. This is the fresh start you need. You are what you eat. If you are well, you can eat for entertainment and for pleasure of the pallette. But if you need to rebuild, you must eat for cellular health. An ill body is made up of trillions of cells, many of whom are dying and malfunctioning. They need your help. Put away the pasta machine and bring in the juicer. The best nourishment in

Prescription for Healing

✷ Cleanse
Wheatgrass
Pure Water
Detoxification
Fasting
Eliminate infestation- –parasites, candida
Wheatgrass implants
Colon Hydrotherapy

✷ Nourish
Wheatgrass
Raw Juice Therapy
Living Foods Diet
Food Combining
Herbs

✷ Rejuvenate
Wheatgrass
Massage, skin brushing
Homeopathy
Exercise–yoga, trampolining, swimming, walking
Acupuncture, Chiropractic
Bio-Magnetics, Ozone

✷ Heal
Rest, Sun–bathing
Deep Breathing, Oxygen
Lifestyle Change
Spiritual, religious study
Personal growth work
Love

the world is provided by enzyme-rich, organic, living foods. If supple ments could do it, then our fast paced society would have replaced the lunch with a pouch full of pills. Grasses, dried or fresh, are complete foods. Ask any of the grazing animals or the guinea pigs who nearly died on a diet of common vegetables *(p. 60)*. It's not hard to find healthy living food. Just spend more time in the garden. Many delicious and healthful foods never make it to the market. Eat sparingly and in proper combinations. As a society we Americans are gluttons, overburdening our bodies with excess that either turns into fat or ferments into a breeding ground for yeasts and parasites. Temperance is a virtue that will add years to your life and life to your years. Add grass, fresh or powdered, to carrot, celery, spinach, beet and other juice combinations. There are an infinite number of recipes available and you are certain to find a combination you love.[5] Explore the world of nutritional herbs and seek ones that apply to your condition.

It's not the food in your life; it's the life in your food.

Step III Rejuvenate

Once the house is clean and the structure is strengthened, it is time to add the accouterments. Get the fireplace working; fix the furniture; replace the appliances. You are out of danger now. The house won't collapse. But it is not fully ready to live in and you can't unwind and let things fall apart again. Although the pressure is off, you continue your cleansing routine of grass, colonics, enemas, implants and living foods diet. Expand your therapy to include regular massage. Start an exercise program appropriate to your condition—be it yoga, walking, swimming, trampolining. Consult with other health professionals who can further your rejuvenation—chiropractors, homeopaths, acupuncturists. Explore different therapies such as oxygen therapies and bio-magnetics. Bathe, brush and prune. It's easier to stay well than to get well.

It's Easier to Stay Well Than to Get Well

Step IV Heal

The ancient sages have long told us that healing is three parts treatment and seven parts nursing. And any homeowner knows, it takes a long time to get things just right. Healing begs for creativity. When it comes right down to it, your lifestyle is either healing or it isn't. Stress is the antithesis of rest. You can't change your health without changing your life. Attitude is healing; environment is healing; laughter is healing; love is healing. All the pieces of the puzzle have to be in place or it won't work. You are out of crisis. Now is the time for spiritual reflection and religious study. Take classes in those things you've always wanted to do for your own growth. In this philosophy of healing, education is medicine; massage is medicine; yoga is medicine.

You can't change your health without changing your life.

Healing is not possible without rest. The fundamentals of health are very basic: oxygen, water, rest. It is amazing, how long you can survive on just those three. This author once had the good fortune of spending two weeks at a Jamaican mineral spa where that was all there was. When left to its own resources, an undamaged body will self-correct. Our lifestyle and our complex modern world get in the way. You don't have to go to a spa. Take baths every day. Use different herbs like ginger or minerals like Epson salt. Bathe in the sun. Brush your skin. Practice breathing. The yogic science of pranayama has many breathing exercises. You will be amazed at how powerful these simple practices are.

Your attitude is so important. In a study of first semester law students under tremendous, unrelenting stress, the students who were optimistic about doing well had more T-cells—natural killer cell activity—than they had before the semester began. The pessimists had no increase in these immune factors. These findings support the notion that thoughts and feelings affect the immune system.[6]

Every disease is both strengthening and weakening. Many who have fought a major health battle are stronger for it. But years of living on the edge of illness is debilitating. Prevention of disease is one of the secrets to longevity. Our society is focused on achieving healing through chemistry. If you have got a headache, take an aspirin. If you have got a stomach-ache take Di-gel. If you have an infection, take an antibiotic. But these are not healers. They are symptom modifiers. Healing is greater than chemistry. If you want to achieve true healing, the restoration of balanced health from disease, include grass as part of a comprehensive wellness program.

The Chlorophyll Cocktail that Cures

Rebuilds the blood	Alkalinizes the blood
Increases hemoglobin production	Neutralizes toxins
Heals wounds	Purges the liver
Cleanses the colon	Stimulates enzyme activity
Anti-bacterial	Chelates out heavy metals

How to Take Wheatgrass

Everyone agrees that finding a convenient way of taking fresh squeezed grass juice is the hardest part of the program. Powdered grasses are partly a response to that. What could be easier than spooning powder into a glass or downing some tablets. We will discuss the differences between the grasses soon, but in this section we will focus on fresh.

Wheatgrass differs from conventional therapies in that it requires more work on the part of the patient. The concepts upon which wheatgrass therapy is built are diametrically opposed to the pill popping style of modern medicine. Grass therapy involves the patient directly. It's hard work but it's impossible to turn that responsibility over to a doctor or anyone else. Like Phillipe Petite, you are taking your life in your own hands. It's hard work. It takes courage and your life will never be the same. You will revamp your kitchen, your diet, your attitudes, your lifestyle, your beliefs. It will change you forever, but you will have your life instead of flirting with death. You will have your health. Here are some ways to make it manageable.

Home Juicing

Juicing at home is the most economical and efficient approach if you are using grass daily. Volume users who juice 3–8 ounces daily for drinking, enemas and implants, are really forced by the economics to grow and juice their own. But growing and juicing take time and effort. In Ann Wigmore's day, there was no other choice and many could not keep up with the regimen. Again, if your life is at stake, you will take the time to clean the juicer five times per day. But today there are other choices. Professional growers provide grass to health stores in every major city across the nation and you can have it fedexed if you live far from town *(see Resources for names and numbers)*. Several professional retreat centers offer three week programs where the grass is grown for you and the juicers are always running. Your local juice bar and health store in all likelihood offers grass juice by the ounce. Today there is a greater variety of juicing machines available then ever before. For details on growing and juicing see *Grow Your Own* and *The Juicers*.

Chewing

Just grab a handful and munch. This is the easiest way to get your grass and it tastes great, too. Here are the advantages and disadvantages. If you don't like the taste of wheatgrass juice, you will probably find chewing more pleasant. Somehow the smaller quantities of chewing and mixing with saliva avoids the shot of strong flavor that comes with a glass of juice. Since every mouthful is approximately ¼ ounce of juice, it can take about 8 mouthfuls to get a 2 ounce serving. That could be a real jawbreaker. But, if you don't mind the exercise, what you spend in time chewing, you save in time setting up, juicing and cleaning your juicing machine. And it's portable! If you pack a zip lock bag with 2 ounces of grass, you can chew it in the car. Since chewing takes time, this is a good place for it. You know that when you finish it, you've downed nearly 2 ounces of juice. Grass is 90% water and thoroughly dry pulp weighs almost nothing.

The mouth is the ultimate juicing machine. While users decry the relative advantages and disadvantages of juicing machines, all agree that the mouth sucks the grass driest and preserves the most enzymes. The biggest disadvantage is that you cannot chew the volume you need for curative results. Eight ounces daily is a typical therapeutic dosage for a serious illness. Two ounces per day is all that can be practically chewed. Not having to deal with juicing, setup and cleanup time is the obvious advantage. But the downside is cleaning up your teeth. Let's just say your new habit will be a boon to the toothpick industry. Grass finds its way into every crevice. It is truly a green floss and chlorophyll is a proven antiseptic and restorer of gum tissue, tightening the gums around the teeth and successfully controlling pyorrhea. But it is also a jaw workout and will take some getting used to. It's just like going to the gym; don't give up on your first try. It gets easier as your muscles develop.

Juice Bars

The next easiest way to get your grass juice is to drive to the nearest juice bar. Juice bars and smoothie bars that serve juices are popping up all over. Companies like Jamba Juice and Zuka Juice *(see Resources)* are establishing hundreds of stores nationwide doing for juice what Starbucks has done for coffee. These stores know there is a juicing revolution. Home juicer appliance sales have exceeded 1 billion dollars. Health food stores often have juice bars inside. If they don't, they may sell fresh grass packaged for you to take home and juice. That saves you the effort of growing it yourself. In 1998, juice bars charged between $1.50 and $2.50 for an ounce of grass juice.

How to Make it Taste Better

If the intensely sweet taste of grass is difficult for you, try this. Grow the grass hydroponically *(see p. 141)*. It tastes milder and all the experts concur that this grass is still potent. Also, try mixing it with other green vegetables. Celery is the best match. Its sodium content nicely balances the sugars in young grass. Other favorites are parsley, alfalfa sprouts, spinach, kale, dandelion, and the sprouts of sunflower, buckwheat and pea shoots. Keep it all green. Add some garlic or ginger, too. You'll find it tastes like a liquid salad and the singular taste of grass is

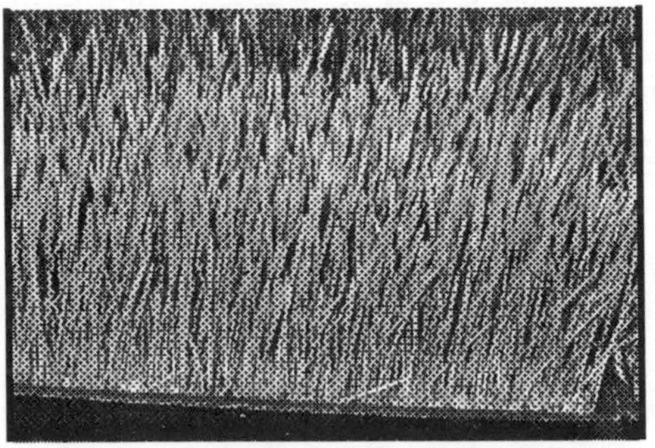

Photo by Robert Ross Optimum Health Inst.

gone. Finally, try growing your grass to the jointing stage. This grass, usually grown outdoors, is more mature and has less of the saccharine sweetness of young grass. *(See Jointing, p. 43.)*

How Much to Drink?

Always drink any kind of grass juice on an empty stomach and then wait 30–45 minutes before drinking or eating anything else. For normal health maintenance, 1–2 ounces of fresh squeezed wheatgrass juice daily is typical. Therapeutic dosages are 4–8 ounces daily. Some take more. Although four ounces can be managed, higher amounts must be taken rectally. Wheatgrass has a strong cleansing effect on the digestive tract. It's practically a green laxative. If you start off taking too much, you will find yourself running to the bathroom. Nausea is also common with over-drinking and it's one of the reasons why therapeutic dosages are taken rectally. This is mollified somewhat when mixing it with the celery and the other green juices previously mentioned. Like anything, once you get used to it, you can take more without effect. It's more than the chlorophyll that does this because drinking bottled alfalfa chlorophyll does not cause diarrhea. Fresh wheatgrass is a high frequency enzyme elixir which jolts your system with a charge that is megavolts above anything else you eat. Even superfoods like blue-green algae can't match the 'chi' of fresh squeezed wheatgrass juice because algae is a powder and can never be fresh juiced. The secret to drinking wheatgrass juice happily is to gradually increase the amount as you become acclimated to it. Raise your dosage one ounce every few days or every week. Drink only what is comfortable. But if you are fighting an illness, no matter what, you will have to take it rectally.

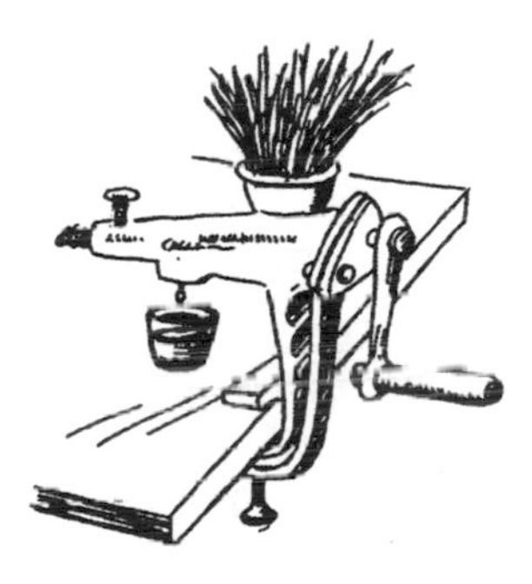

How to Take Wheatgrass Enemas & Implants

Ann Wigmore invented the use of wheatgrass implants. She realized the limits of drinking the juice which can cause nausea in volume. And she knew that only higher quantities would provide therapeutic results. In addition, she believed that the large intestine was not designed to handle the Standard American Diet (SAD) of processed foods and fiberless flour products. Over a period of years, layers of paste build up on the colon walls, shrinking the tunnel diameter and reducing the efficiency of elimination. This is one of the causes of toxemia from which numerous aliments develop. Colonics and enemas assist in reducing the accumulation and restoring normal colon function. She started adding wheatgrass juice to the enemas which helped heal ulcers, sooth the tissues, oppose bad bacteria and nourish the bloodstream. Since enema water is usually expelled immediately, she decided to first cleanse the colon with a wheatgrass enema. Coffee enemas were also used because of their ability to stimulate the liver. In a two quart enema bag, only one ounce of wheatgrass juice was added.

Once the colon was clear, from one or more enemas, a larger dose of wheatgrass was 'implanted' and held for 10–30 minutes or longer. Implants offer a way to introduce into your system 8+ ounces of grass. This is the therapeutic dosage required to purge the liver, purify the bloodstream and detoxify the colon. These three effects are the primary benefits of wheatgrass. Anything that can improve the performance of the liver, cleanse and nourish the bloodstream and accelerate elimination must have a powerful, positive impact on health.

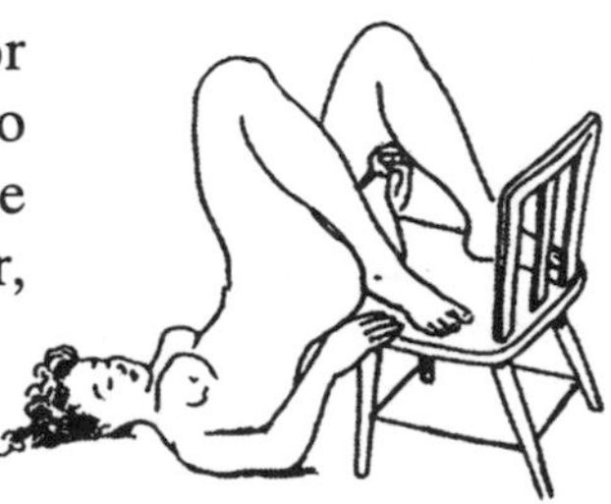

There are a few approaches to taking implants. For one, you could simply increase the dosage of grass in your enema bag. This approach is more wasteful of grass because much of it is flushed out and not retained. But the final round could have a higher dosage and is likely to be retained longer.

The standard approach to an implant uses a bulb syringe. First prepare the colon by cleansing it with a plain water or wheatgrass-water enema described above. It may take multiple enema bags to clear the intestine. A cleared colon is free of solid material in the waste stream and ideally, light in color. Now it is intermission time. Take a break to juice your wheatgrass so it is fresh for the implant. Pour it into a cup. Take a 2–4 ounce bulb syringe, available at drugstores, squeeze out all the air, then suck up all the wheatgrass juice in the cup. Grease the rectum with K-Y jelly or a water soluble lubricant. Either elevate your legs or lay on your left side. Squeeze the bulb slightly to eliminate any air bubbles; Insert and slowly squeeze. Rest for several minutes until you feel capable of taking more. Retention is more important than volume. Some people can hold their grass for hours. The longer the better. Small amounts may even be absorbed. Volume will be dependent on experience. Daily application and gradual increases are the key. Although your first experience may be unappealing, you will develop the skills you need to enema and implant successfully. The benefits are worth the trouble.

A few alternatives: If there is colonic hydrotherapy available, it is preferred. Colonics are more efficient than enemas and more thorough. They are also easier on the patient. You can accomplish more in one colonic than in several enemas. They're easier, faster and more thorough. Another way to implant the wheatgrass is through an extension tube. This is a 12–16 inch enema catheter available by special order from your druggist or through the wheatgrass clinics listed in the *Retreats* chapter. The catheters enable deeper delivery into the colon rather than just into the rectum. The catheter should be lubricated and inserted first. Then insert the tip of the syringe into the catheter end and squeeze.

Depending on your background, this may all sound very outrageous, terribly primitive or at least inelegant. But when you are fighting for your life, issues of dignity and grace go right out the window. This works and in the end, it all comes down to choices. After all, chemotherapy is no picnic.

What About Powders & Tablets? Dried vs. Fresh

The whole process of growing and juicing grass is made infinitely convenient with powdered grass and tablets. This was the theory behind Yoshihide Hagiwara's development of barley grass juice powder. He wanted to make the benefits of grass as readily available to the masses as instant coffee. He succeeded. But while field grown grass is nutritionally superior to grass grown indoors in trays and has prodigious nutritional value, it is generally the Ann Wigmore concept of fresh squeezed juice that is considered the premier grass for healing. Why?

Glandulars are dehydrated animal glands and organs that are powdered and bottled. These supplements supply the complete nutritional needs of the specific organ or gland. (Vegetarians read labels carefully.) If one could carefully dehydrate and grind a body into powder and test the results, it would contain a concentrated source of all the nutrients necessary for human life. However, in spite of the fact that all the elements are there, it won't walk or talk and bears little relation to the original product. While nutrients are necessary to support life, they cannot by themselves recreate it. When you reconstitute a dried tomato, it neither looks nor taste like a fresh tomato. But if you measured its elements, you would find that as tomatoes go, it is nutritionally well endowed. So what is the missing link? What does the nutritionally inferior indoor grown grass have that the superior field and sunshine grown dried grass lacks?

When taking fresh squeezed wheatgrass juice, people say: "I can feel it running through me." Or, "It makes the hair on the back of my neck stand up." The street kids call it a 'buzz.' They feel high. What is that? We call it energy—the life force. The Chinese call it 'Chi' or 'Tao.' The yogis'—'kundalini.' Yoda said: "May the force be with you!" It is the electricity of vibrating, bubbling, scintillating life. Inside the juice, its photons, protons, electrons and quarks are dancing to the music of the cosmos. Our living cells reach out with an irresistible magnetism for these charged nutrients. This magical bonding occurs at an intensity only achievable between living cells. When we are ill and depleted, our cells operate at low intensity, like a weak battery. Only other living cells can provide the electricity we need to recharge. Although there is enzyme activity in dried grass, it is not at the same level of intensity as the fresh. When you really need to jump start a sick body, there is no substitute. If you are trying to heal and rejuvenate, you want the best. Juice the fresh grass; drink in the liquid sunshine. "It's the Real Thing."

Nonetheless, all the research done on grass has been on dried grass. The nutritional value of grass is so prodigious that even the dried powder has powerful effects. These results are measurable and striking *(see Research)*. So it is understandable that there is confusion about this choice. The following two paragraphs are comments by Piter U. Caizer, the

Wheatgrass Messiah, *(see p. 13)* which propose the best of both worlds is to fresh squeeze the outdoor grown grass.

"Tray grown wheatgrass is so intense. What a strong aftertaste. You can drink much more of field grown grass. I prefer to take the wild grass that mother nature has grown in the fields, to the tray grown. There is definitely a big difference between the stuff that is grown in the tray by man, and the stuff that is grown in the field by nature."

"It's powder. Anything that is dried has an incredible loss of vital life force. Even if you dry it carefully. Dried is dormant. Dormant is dormant. Problem with all that dried stuff is the life force–the ethereal energy is dissipated and lost. You're a living, vibrating entity. So when you drink something highly vibrational like grass—grass juice is the highest vibration—it tunes right into you and brings up your vibration."

More and more Americans want to juice and are advised to do so, but the truth is they don't. For those people who don't have time to juice and don't have a health emergency, powdered-tableted grass is a convenient, portable and palatable way to get more green foods into the diet. The therapeutic dosage for powders is much higher than what is recommended on the bottle. While not a panacea, grass is a potent natural medicine in any of its forms: barley, Kamut, whole leaf dried, powdered juice or fresh squeezed. But as with any herb, its healing power increases with its freshness. Your decision about 'which grass,' should be made according to your degree of wellness/illness.

We're a society of fast foods, we want things right away. The powdered grasses all work and they all don't work... Yes, scientifically you can compare the fresh and the powdered, but there is no way that you can tell me that the powdered is even a close equivalent to the fresh. You've got living, vibrating, energy —even breathing the stuff in is powerful. Yes, the alternatives are convenient, but how sick are you? If my life depended on it, I would want optimum results in the shortest period of time. As far as I'm concerned, for that, you've got to go with the fresh.
—Michael Bergonzi, Optimum Health Institute in San Diego

Allergies and Sensitivities

Many people ask: If I have an allergy to wheat, can I still take wheatgrass? People who have allergic responses to wheat and wheat products are usually reacting to gluten, the sticky protein found in the grains of wheat, barley and rye. This is the same 'glue' that is in plaster of Paris. The overconsumption of flour products in the American diet has overburdened our systems, forcing us to rebel with our 'allergic' response. Its paste has also plugged up our intestines. Wheat grass is different than wheat. One is a grain, the other a green vegetable. The green vegetable grass contains no gluten. It is no more 'allergic' than spinach, kale, chard or lettuce. In fact, it contains anti-allergic factors. Since allergies are immune responses to toxic irritants, detoxification is crucial to any allergy treatment program. As discussed, colon health is key. In addition to the liver purging, blood purifying and oxygenating capacity of grass, it coats the colon tissues with soothing, anti-bacteriostatic chlorophyll. Whole leaf wheatgrass powder also provides a high quality vegetable fiber—twice the fiber of bran—that maintains regularity. Add this powder to a daily juice to create a fiber rich health drink.

Even if grass did have gluten in it—which it definitely has not—you could always switch to barley grass or Kamut. Barley grain has less gluten than wheat grain and Kamut grain has a different kind of gluten. The International Food Allergy Association found Kamut flour was okay "for most wheat sensitive people."

First Aid

You do not have to be sick to use wheatgrass. Grass has numerous first aid applications from fatigue and sleeplessness to athlete's foot, bad breath, body odor and burns.

When taken full strength and strained through a paper filter, wheat grass juice has a variety of applications for the eyes, ears, nose and throat. Use only one or two drops in the eyes for eye strain and tension. Slight stinging is normal, but only momentary. Dr Gary Hall, medical director of the Eye Surgery Institute in Phoenix, Arizona recommends wheatgrass juice for anyone who shows signs of retinal disturbances or has a history of macular degeneration. Because of its richness in magnesium, wheatgrass acts as a smooth muscle relaxant which may be the reason glaucoma patients report relief.

Placed in the nose, one or two filtered drops reduces inflamed nasal passages and soothes mucous membranes irritated from allergies. Try it for stuffy nose, sinusitis, rhinitis, bronchitis, itchy palate, etc. It easily breaks through nasal congestion and doesn't have the boomerang effect of drugs that leave you with even greater congestion after they wear off. In 1941, Drs. Redpath and Davis, Eye, Ear, Nose and Throat specialists at Temple University, treated over a thousand cases of allergy and upper respiratory problems with chlorophyll and reported impressive results[7]. Gargle with the unfiltered juice at the first sign of a cold or sore throat. Wheatgrass' bacteriostatic and antiseptic action provide genuine relief. It is also a great mouthwash and leaves your breath smelling fresh even after eating garlic. Bleeding gums, trench mouth, gingivitis, and periodontal (gum) problems in general, are very responsive to wheatgrass.

Wheatgrass is great to have around the house for cuts, bruises, rashes, burns and bangs. Make a bandage from gauze dipped in wheatgrass juice. Even better, re-dip the pulp and put a little under the bandage. If it is a large wound, wrap it in soaked gauze or pulp and protect it with a towel to prevent dripping. The American Journal of Surgery reported that in experiments with over 1,000 surgically wounded animals, chlorophyll increased the rate of healing by 25% over the non-chlorophyll control group.[8] Until you use it, it is hard to appreciate just how rapidly it reduces swellings, takes the sting out of burns and heals wounds, frequently without leaving scars. You can also use grass juice powder, either wheat or barley or Kamut, for first aid. Just dip your moist gauze into the powder.

And now for the cosmetics. What a great facial! Just dip some moist gauze into the powder, rub it on and let it sit. You'll look green, but you'll feel great and have fun scaring your family. It's also a great skin cleanser, perfect for acne and black heads. There are documented results with skin problems like eczema (*see Research*), and it's been used for itchy skin, poison ivy, sores, boils, cuts, burns, insect bites and dandruff. Many skin conditions are related to liver and colon congestion and cellular hyperacidity, so in addition to topical application, you must drink the juice and get on a total health program for a long range solution.

Other reported uses of grass juice are: Hemorrhoids—make cotton-grass suppositories. Asthma and bronchitis—put compresses of wet pulp on the back and chest and drink the juice because it is an expectorant.

Weight reduction—a grass juice cocktail before meals stays the appetite. Some users also swear a cocktail before bedtime helps them sleep.

Wheatgrass the New Miracle Drug?

Don't expect to find wheatgrass on your doctor's prescription pad. It costs 300–400 million dollars to evaluate a drug for FDA approval. Since grass is hard to patent, drug companies are not likely to make that investment. The basic approach of modern medicine is to treat physical abnormalities with chemistry and surgery. Wheatgrass does not fit into this model. Fresh wheatgrass juice enlivens the spiritual in addition to nourishing the physical. It's an electrical elixir transmitting a charge of energy that arcs across neurons and nerve fibers, revitalizing trillions of cells. From the conventional medical standpoint, it's quackery. Nonetheless, the medical establishment has been forced to expand their perspective before on such mysterious therapies as acupuncture. No, wheatgrass is not a magic potion. But when used as part of a total health revitalization program with an indefatigable commitment to wellness, it is a powerful healing agent which accelerates and maximizes your potential. (For more, see *Science & Wheatgrass.*)

> *The cure of many diseases is unknown to the physicians...because they are ignorant of the whole, which ought to be studied also; for the part can never be well unless the whole is well... [This] is the great error of our day in the treatment of the human body, that the physicians separate the soul from the body.* —Plato, 477–347 BC.

Real Stories from Real People

With my fitness level, my drive and desire, I'm not going to lose. I can't lose. —Lance Armstrong, USA Cycling champion on his battle against testicular cancer

The testimonials in this chapter cannot just be dismissed as anecdotal evidence. These are real peoples' lives. There is a new branch of study called evidence-based medicine which considers everybody an experiment of one. The stories herein represent only a fraction of the thousands available. Comments inside [ ] brackets are the author's.

Breast Cancer

—Anne-Marie Baker, Ft. Meyers, Florida

On March 1 of 1996 I had a needle biopsy which, when the results came in, confirmed breast cancer. My doctor recommended an immediate radical mastectomy and a full program of chemotherapy and radiation. I'm a registered nurse. As part of my job, I am in and out of hospitals and doctors offices, all the time. So I am well aware of the effects of cancer and the results of the treatment. But I just could not go through with it. So I went against doctor's orders. My colleagues were mortified. First, they were truly concerned for my well being and wondered if I was mentally all there. Secondly, they didn't understand how I could go against the system in which I worked. Some said that I was afraid of how I would look, but that had nothing to do with it. I have seen people go through this therapy hundreds of times and I just don't feel in my heart of hearts that it is the right choice for me. I would never steer anyone away from conventional treatment if that's what they choose. I respect everybody's choice and it is my job to support them in that choice. But for me, I just could not go through with it. Philosophically, I feel it's the wrong approach—destroying all the good cells along with the cancer cells. I want to strengthen my immune system, not debilitate it. After all, if the immune system was strong enough in the first place, the cancer would not have grown. Depressing it more makes it an even harder fight.

So, I was recommended to a holistic doctor who was actually a chiropractor with an extended practice. He worked closely with me in developing a nutritional program that included wheatgrass, raw vegetable juices, supplements, exercise, and detox. For a few months, I was really bombarding this thing pretty intensively. I started out with 1–2 ounces of grass juice, then gradually built up to 6–9 oz. [orally]. I was also doing lots of cleansing through diet, enemas and colon hydrotherapy. After every colonic, my therapist would put in 4 oz. of wheatgrass juice. I would hold that for several hours and I'd say about half of it was absorbed. Then I would do my own enemas daily along with the with 4 oz. implants using a bulb syringe. And of course, lots of raw juices during the day—at least 4 juices of 8–12 oz. each. I juiced everything, lots of greens, lots of garlic, sprouts, mostly organic veggies except when I couldn't get them.

Most of my friends are physicians and nurses and at first they told me 'you're in denial,' but now they're hushed. You know since that time, I never called in sick. Not once! I have a better attendance record than anyone in my office.

Since I started the wheatgrass I have more energy than when I was cheerleading in highschool! It's holistic, of course. Exercise is a big part of it, too and I do lots of yoga. Now I'm on a regular routine at a more relaxed pace. I'm working a full time job with a lot of running around. I had an AMASS test six months ago and it came up negative. No sign of cancer, anywhere! I have a lot of confidence in what I'm doing. I think it's important to surround yourself with positive supportive people. Otherwise you're fighting a double fight, battling with the outer world as well as the inner one. You just don't need to raise the odds. Besides, you learn a lot from supportive people. Everybody you meet either gives you a boost or a knock. You've got to change your lifestyle. I'm glad I have the flexibility to run home and make a juice. My health is more important than my job. If you don't change the [lifestyle] circumstances which led to the problem then I don't care if you cut it, burn it, or poison it, it will come back.

I don't think there is one answer to cancer or health. It's an accumulation of a whole bunch different therapies. But wheatgrass was fundamental to my therapy. I read that it's the abscisic acid in wheatgrass that targets the cancer cells. [A plant hormone that regulates growth and metabolism.] I just wish I could sneak a juicer into the hospital and make wheatgrass for all my patients.

Bladder Cancer

—Dorothy Naylor, Naples, Florida

What can I say, I'm supposed to be dead according to them. My bladder was totally covered with tumors. Now, my MRI doesn't show any. None. My bladder was cut so many times and lasered and scraped and fried. Then the chemotherapy, the radiation, all the drugs. It was awful. It fried the surface of my bladder. And during that whole time, whenever I would go back for an examination, there would always be another tumor or two. The MRI and sonogram always showed up some new cancers. I had them removed at least six times. He was always scraping them out and cutting them out. So the doctor says, we can't do this forever, let's remove it. He wants to remove the bladder. That's his alternative. Surgery. Then I'd have a tube and a bag hanging out....it's terrible. But I didn't have any alternatives, at that time. So I went into the operating room and right there, just before they were going to take it out, I had a heart attack on the table. It wasn't severe, but it was enough to stop the surgery.

That heart attack saved my life. You see the doctor told me and my family that I wouldn't last long enough to visit my children in August. This was April. My son was furious with him. I didn't feel like I was going to die. It was at that point, I decided to look for alternatives. I wasn't going to let them remove it and no more laser or scraping the tumors, either. I had enough of that.

My daughter from Germany got me started on wheatgrass. No one else in the family knew about it but her. I have 13 children, you know. With her help, I started drinking wheatgrass everyday. I was religious about it. My children got me a Green Power juicer. I like it but it's a lot to clean. Then I started growing my own wheatgrass, but I just couldn't do that. That was too hard. Now, I get it right down the street. Back then, I got it shipped in for $8/lb. At one point I took as much as 7 ounces of grass juice per day. But then, my body couldn't tolerate it, so now I'm back down to 6 ounces per day. Two ounces in the morning, 2 in the afternoon and 2 in the evening.

It's been 4 years now, and I only had one tumor since because I was visiting my daughter up by you and I was neglectful. I went off the wheatgrass. When I got home I had a tumor! Well, I've never left the program since and there has never been another tumor. My MRI and sonogram are clean. My doctor says: "Just keep doing what you're doing." He's a nice doctor but strictly conventional.

My bladder is not a problem anymore; now I'm dealing with my asthma. Charles is giving me ozone treatments. I take it through the ear and in one minute it goes through the entire body. It's fantastic. My asthma clears right up. It's very big in Germany. These people at the Optimum Institute here in Naples are wonderful. They are not in it for the money. They don't charge for the ozone. I have to apply it myself—legal reasons. Wonderful things are happening down here.

Candidiasis, Irritable Bowel, Leaky Gut Syndrome

Tony Gentile of Malaga NJ is a former collegiate wrestler and teacher who had to quit his job because of complications from irritable bowel and leaky gut syndrome.

My problems started during my wrestling years. Because we had to meet the weight requirements I was on a fiberless diet because fiber adds weight. So I'd have two candy bars for dinner and get false energy from that instead of eating real food. I also got dehydrated because water was a no-no—it adds weight. The competition years ruined my health by starving my body for nutrition and killing my intestines without fiber or water. I had chronic constipation and dehydration and became dependent on laxatives. That was the start of my digestion problems. From there I developed candidiasis, leaky gut syndrome and eventually arthritis. My knees got so bad I had two operations on them and they are still [many years later] weak and painful. My intestinal condition was so debilitating I couldn't focus at work. I quit my job.

I've been fighting this fight for 20 years, now. I've tried everything; nothing works. Only wheatgrass has helped me. The implants are great. I do an enema first to clear myself out and to better hold the implant. I take 6 ounces of [wheatgrass] juice mixed with about 2 ounces of aloe vera gel and a little warm water. Making it body temperature helps retain it better. I put it [all] in an enema bag and attach the 12 inch tube. [A 12 inch enema extension catheter is available by special order from your pharmacy.] I get on the slant board with feet high to get it as far in as possible so it reaches the portal vein and goes to the liver. I let it slowly leak in, real slow and hold it for as long as possible, ideally 30 minutes. In 2 minutes the burning [irritable bowel] is gone and I'm feeling high. Since

I've been doing wheatgrass, my digestion is ten times better. My hair got thicker, the white spots left my nails, the dark rings around my eyes cleared up and the pain in my legs is gone. There is a change in my muscle quality, too. They're firmer; my knees last much longer.

After 20 days of implants, 6 ounces each twice per day with the extension tube, that fire [of pain] in my colon was gone. I mean gone. My liver is not swelled up anymore. That hardness, it's not there, gone. Not sure why, but you're getting really high assimilation of a complete food. Then you have the implants purging the liver and cleaning the blood. I can feel it. The blood feeds the muscles, ligaments and tendons. It's the chlorophyll. It's just healing and cleansing. Did you read that chlorophyll book by [Dr. Bernard] Jensen? It's the enzymes, too. They're catalysts. My digestion is 100% better. I take a couple of shots before [15-20 minutes] I eat. It gets my cylinders firing. My irritable bowel...it's the only thing that has helped me. I'm growing tons of sunflower and buckwheat [sprouts] now, too. I make lots of blended drinks [with them]. My candida and my digestion are 100% better. I sleep better at night, too....less of that AM toxicity when I wake up.

—

Wheatgrass contains raw chlorophyll. Chlorophyll is condensed sunlight. Since we are light beings, spirit and soul inside solid bodies, the light force vibrates through the physical body. That's the energy you feel. Wheatgrass is a spiritual food. It nourishes you on a spiritual level as well as the physical.

—Piter U. Caizer, The Wheatgrass Messiah

Grass and Marathon Endurance

—Bruce McVay, Salt Lake City, Utah[1]

The newspapers have recently published several articles on animal studies that have proved that very low calorie, highly nutritional diets enhance strength, endurance and longevity. Wheatgrass has by lab analysis proven to be very close to, if not the foremost plant in the category of nutritive value per ounce. Being an ultra–distance runner for 5 years [races from 30 to 100 miles or more], I tried every physically enhancing supplement and natural food source to improve my health, stamina, and oxygenation process. One and one-half years ago, I began juicing 1–6 ounces of grass daily. After 1 to 2 months of cleansing—which consisted of a few headaches and itching—I began to notice tremendous changes in my performance, energy, stamina, etc. Always in the past, I was a middle to rear of the pack racer. Wheatgrass changed that.

This year, I ran a 26 mile race in the Teton Mountains and won by half an hour. This race was in the top of the Tetons, 12,000 to 13,000 feet high, and down to Jackson Hole, Wyoming. Ultra distance runners from all over the United States participated. At the 18 mile checkpoint, some people with twice the amount of training as I, quit the race. Two or three finished. Something I noticed was that I never hit the wall (nausea, etc.) like I had sometimes in previous races. I was exhilarated all the way to the end of the race.

On June 5, I participated in an 11.5 hour run-a-thon at the Utah State Prison. I ran over 62 miles eating primarily wheat and barley grass mixed with fresh apple juice. During this time, I expended over 6,500 calories. My nearest competitor was 10 miles behind. These are factual examples, but the best result is the way I feel all the time and during the race—very comfortable under physical and mental stress.

Some people would think that I increased my training or trained long and hard for these races, but in reality, I cut my training in half during this period of time. I had cut my training mileage from 60–90 miles per week to 30–40 miles per week. Imagine less training, but better performance!

I have broken most of my own personal records. Another thing that amazed me about the prison race, was that my laps at the last of the race were as fast as the laps at the beginning. Wheatgrass helps numerous processes in the body, but improved oxygenation [enhanced oxygen absorption] is the best one. I believe in it.

Colon, Lymph and Liver Cancer

—Gary Garrett, Gainesville, Florida

In June 1995 I was told that I had a large tumor in my colon. Dr. Wyshbaum my oncologist, told me I had only one option: have immediate surgery to remove a section of the colon. I followed this advice and had the surgery.

The doctor then told me the tumor was approximately the size of a baseball and had grown through the colon wall to adjoining tissue and the cancer had metastasized to six lymph nodes and my liver. The prognosis was not very good. They expected further tumor activity within one year. My surgeon strongly recommended that I have surgery on the liver and begin a course of radiation of the lymph system and chemotherapy for the colon as soon as possible. Before I could even consider this I began to have complications from the surgery: the incision would not close up. It required new dressings three times a day for the 18 days I was hospitalized, and for months after I was sent home. As long as my wound would not heal I could not have any toxic chemo or radiation. After surgery I could not have anything by mouth–even water–for 12 days, only IV's.

It was during this time that a good friend found out I was sick, contacted my wife and offered his assistance. He had been fighting a similar battle with cancer for some time. He has read many books and done much research on his own. He gained valuable information which he was willing to share with me. This included diet changes and learning all about something called wheatgrass. As I mentioned earlier, my wound required dressing three times daily. Even with strong pain killers like morphine it was almost unbearable to have the nurses wrap and unwrap the area so often. After 12 days on IV's I could finally begin to take liquids by mouth. I first experienced wheatgrass juice when a friend brought 6 ounces to my hospital room, which I drank right down. I didn't notice anything until the next day when my morning nurse came to dress my wound. As I mentioned, it became so painful I dreaded these times. What a pleasant surprise the next day when my nurse began to take the dressing off and I could tolerate it because it was not as sensitive as before. Then an amazing thing happened, the nurse pulled out the packing and we both noticed it had turned green.

At first this scared her because they are trained to watch for changes in the dressings. When they are green it usually means a very serious infection. However, I had told the nurse about the wheatgrass and she recognized that the color change was a result of the chlorophyll. Imagine my surprise, in one day, to drink crude chlorophyll which is dark green which passed through my entire system into my blood stream to find its way to this external wound and reduce the pain and speed up the healing process.

As time went by my home care nurses were amazed at how quickly my wound healed. The wound packing continued to have a green tint throughout the whole time. When I left the hospital I was so weak I could hardly climb a flight of stairs. But by taking wheatgrass juice daily, in a matter of weeks, I was able to return to a full work schedule with no new tumor growth in my colon.

The following was written by Mrs. Gary (Kathleen) Garrett, July 1998.

Gary had colon cancer and six lymph cancers and one spot on the liver. Dr. Botonay heavily insisted on chemo and radiation. I asked him if we did all that would we get rid of the cancer? He said it would give Gary six more months. Gary and I left his office determined to seek alternatives. Why poison the immune system if you're trying to strengthen it? I'm an X-ray technician and Mom is a nurse. We saw what wheatgrass was doing for Gary even though we never had a drop of it. I thought it was a joke at first; now I'm a total believer. Half of his incision reopened and we watched the grass heal it. That was amazing and wheatgrass kept the cancer out, too. It never came back to the colon; it never came back to the lymph and the liver tumor had completely calcified. Gary's C.E.A. was only at 40. It should have been in the hundreds or thousands with cancer as advanced as his. His was incredibly low. We had the cancer beat. Even the bone scans and other tests proved that it never went into bone or the lungs or anywhere else. But the calcified remains of the tumor on the liver obstructed his bile so they put a tube in his side to drain the bile. That caused a bad infection. He was hospitalized sixteen times between April and December. They kept redoing the tube and reinfecting the liver and the infection traveled throughout his whole body. His body started to shut down. He wasn't able to make his own platelets. A normal platelet count is 150,000. Gary's was at 61,000. You start hemorrhaging when you're that low. We don't take blood transfusions; we're Jehovah Witness'. We needed to raise the platelets without a transfusion. All our hopes turned to wheatgrass. Dr. Smith [cancer surgeon] permitted us to give 2 oz of wheatgrass every 4 hours through a tube down his nose that went directly into his small intestine. My son Kenny and I grew it, juiced it and gave it to him. Everyday we watched his platelet count rise. There was no question about it. They took a blood work up every day. It's fully documented. But Gary was in septic shock from the infection. He was too ill even to raise a fever.

Gary's platelets rose every day for seven days. From 61,000 to 141,000 strictly from the wheatgrass, nothing else. It's all documented through the lab work. How could someone that ill, with his immune system and his kidneys both shut down, make such a comeback nearly reaching normal blood count? Dr. Smith called it a medical phenomenon. Now, he's taking wheatgrass!

Gary didn't make it. But he lived 3 years despite Dr. Wyshbaum's prognosis of 6 months. Gary had a friend in Pennsylvania with colon cancer and was also given six months. He went through the recommended chemo and radiation and died in 6 months anyway. Gary spent 3 years on wheatgrass working 16 hour days. Despite his excess weight—he was over 300 lbs.—Gary was running up and down stairs and in between he ran back to the hospital to get intravenous antibiotics for the infection they induced. His energy was remarkable. He was back to work 10 times faster than they predicted. He took 4–8 ounces of juice every day, both orally and by implants. He held those implants in all day sometimes. We went to another hospital for five weeks and I didn't have any wheatgrass there. His immune system wasn't able to counter the toxic overload. He was getting thinner and weaker. It was the double whammy of that hospital induced liver infection in a critical organ and the toxic overload that overcame his immune system. If I only had wheatgrass during those five weeks away from home, it could have made all the difference.

In Gary's honor, his wife Kathleen and sons continue to grow wheatgrass for people in their area and beyond. (See Resources: Wheatgrass Express.)

Senior Citizen on Grass Out-Lifts Bodybuilders

—Leland Bender, Bloomingdale, Illinois.

My son Eddie at age 51 was ready to die. He smoked for 30 years—-3–4 packs every day since he was 18. He weighed 315 pounds and couldn't walk a hundred feet without gasping for breath. Doc said: 'you got plugged arteries.' He was only 50 years old! But he was not ready to die. I got him on the wheatgrass program. Steve, listen to me—you cannot believe the turnaround in him. He did wheatgrass juice every day; he lost 50 lbs; he dropped the meat 100%. He's eating sprouts. Evidently wheatgrass does rejuvenate the lungs, because now he can trot right along with the rest of us. He's feeling great.

And I'll tell you what, he's not going back to work. Not the same work. We're going into the wheatgrass business. Father and son. We're calling it NutriGrow. We're going to supply gyms and individuals with the growing materials via network marketing. It's gonna explode. It's got to.

You wanna hear a story? Call Addison Gold's gym [Bloomingdale, IL]; speak to Mike Niewinski, the manager. Ask him about wheatgrass. He'll tell you. He won't be without it. We can't grow it fast enough for him. I got the grow racks in there with lights and automatic misters. We sold over 300 trays of grass and 50 juicers. The guys in the gym are going ape over the stuff. They don't need steroids anymore. They feel an energy surge within 15–20 minutes. The bloodstream just sucks that nutrition right in. Just sucks it in. They lift another 20%. That's the edge on the competition. I'm telling you, it's going to explode. And they need your sprouts, too. Pea sprouts, broccoli sprouts, sunflower, buckwheat. Your sprout bags are great.

It's just amazing stuff. In the past year I've walked in excess of 700 miles across Illinois. I do 2–3 miles per day everyday in all weather. I'm telling you it's the wheatgrass. I can't believe my strength. I'll be seventy-two on November 24th. I've had cardiovascular strength before, but now...I go on the spin cycle for an hour without stopping. There's an instructor putting us through the paces. The young ones are dropping out after 20 minutes. But I roll through a full solid hour without gasping for breath. And my strength! My trainer will tell you, I out-lift in endurance 70% of the gym regardless of their age. And we work hard. One and a half, two hours, I mean the perspiration just rolls off of you. And I don't have a sore spot on my body. I'm as flexible now as when I was 18. And my weight—I lost 40 lbs. I'm 235 now I was 275. It definitely eliminates the hunger.

Yeah, I tried the powder. You don't get the same effect. Common sense will tell you, it's gotta be live. You can feel an energy surge within 15 minutes. I grow my grass usually 5–6 inches for 8 or so days. It's a real rich green. I'm up to 6 ounces per day, now. [Drinking] Three in the morning; three in the evening. No antacids any more, whatsoever. Chlorophyll—that's the best antacid.

You know, I've had a kind of gravel in my voice for years. I chalked it up to age. People over 50 have a lot of phlegm. After 2 weeks on wheatgrass I spit up plugs of phlegm. Gone. it's gone! You can hear my voice echoing. I told the doc, he say's no wonder, it's all clean in there.

I've been through the hospital system. I had advanced gout. Too much meat. I mean this is the old days, of course. Now my former hematologist is on alfalfa tablets and is about to come into the wheatgrass program. He actually lost weight. It definitely takes care of that gnawing feeling of hunger. You're empty, but satisfied.

And there's another miracle, too. Mike's father-in-law—his fingers were always tingling and then they would go numb. This was going on for 2 solid years. Mike walks into the gym and he says: 'Lee, you cured my father-in-law. No more pain; no more tingling in his fingers.' It's two months later, still no numbing, no tingling, no pain.

I'll tell you what. After they get rid of the cigarettes, they should outlaw McDonald's and Burger King. I mean, how many generations do we have to sicken before they figure it out. We can't afford to pay for it anymore. The insurance companies are way out of line. The people can't pay. The sick are just going to wind up on public aid. That's the way it's gonna be. And that goes for the nursing homes, too. Half the people in there didn't have to be if they had just done a better job on their exercise and nutrition. I'll tell you where the problem begins, it's the kids. Mommy I want french fries; I want pretzels...chocolate. Only 1% of the kids have good nutrition. They get away with it because of the TV. It hypnotizes them. I'll tell you what, we gotta get this stuff on Oprah Winfrey. It's all about the media. That's all it will take. But one way or the other, the word will come out. It has to. This stuff is too incredible.

—

Never Underestimate the Power of Nature

Grass and Melanoma

—Neva Whetzel, Singers Glen, Virginia

I noticed a small mole had begun to grow. It was about 1/4 inch. I checked it out with my kinesiologist who is really great and we determined it was cancerous. I went to see a specialist who diagnosed it as a melanoma and gave me a "one in a million" chance of survival. He was adamant that it be surgically removed including all surrounding tissue. He wanted to remove a 3x5 area going down into the muscle and also removing my lymph nodes. It is was in the upper thigh area. Then he would graft skin back on. He wanted me on a full chemotherapy and radiation program. This was only a 1/4 inch melanoma mind you. The only thing I would let him do was remove the quarter inch spot. No more. Then I went on the wheatgrass.

I have been eating well all along—I am not a junk food person. So I went back to my kinesiologist and we checked out what to do for it and that's when I started the wheatgrass. Actually any kind of grass checked out good, barley, Kamut, even the powdered grasses tested fine. I took those whenever I was without the fresh. Several herbs checked out good as well and I took those too. I did lots of things. You can't say it was just the wheatgrass.

I started growing the fresh grass in my backyard in the spring. I grew it in the ground and let it mature [to the jointing stage]. I would fast for 10 days at a time and repeat that a few times. I drank 8 ounces at a time—I'm an all or nothing person. Sometimes it would nauseate me and I'd get dizzy from it—probably a liver reaction from dumping toxins. But we had a lot of wheatgrass and we wanted to use it up because it's no good unless it's fresh. So my husband Dennis drank it, too. We both took 8 oz of wheatgrass on an empty stomach over the course of one hour. That's how I took it. One thing I noticed after a while, my hair softened. It just got soft, like a baby's hair. My facial skin felt soft, too. What a reaction. I couldn't figure out where it was coming from unless it was the grass. But what really amazed me was that Dennis had the same response. His hair was silkier, too! And it wasn't like we even told each other about it. It just happened.

Well, I went back to the same doctor a year and a half later and you know what he said: 'I was really worried about you. Apparently what we did for you in the office that day cured you.'

Slipped Disk Dog, Lyme Disease, Emotions

— Loretta Vainius, Malvern Pennsylvania.

My dog Spirgis is a long haired mini dachshund. When he was only 4 years old, he got really, really sick. He had a high fever; he could not move or do anything. He was defecating wherever he sat. His eyes were murky. He was definitely dying. I took him to the vet. They gave him steroids and antibiotics. A friend recommended a veterinary chiropractor who discovered Spirgis had a slipped disk. He gave him an adjustment but when he tested him by pinching him between the toes, there was no feeling. Spirgis didn't respond at all. The doctor said that it had gone too far and there was nothing he could do. He suggested a veterinary surgeon. So I went to this veterinary surgeon and I said: Is there anything you can do, he's got a slipped disk and he's paralyzed. The surgeon says: there is a possibility we might be able to do something, but it's not very certain. He told me the operation would cost $1,800 and there was no guarantee. He said: if he still can't walk they would make a little wheel for him to roll on.

Well, it took me a while to learn to use my own therapy [she uses wheatgrass and gives it to others]. I mean, how do you give a dog an enema? I figured, what did I have to lose? It was a good thing it was summer! I gave him 4 enemas in five days. We did it outside on a table. It took three of us to give the dog an enema. First I gave him a coffee enema followed by a tray of grass [8-10 ounces]. After the fourth implant he shifted his body. He couldn't go anywhere of course but it was an indication that he was responding. Then I got an eyedropper and held his snoot and I put 2 ounces of grass right into his mouth. Two in the morning and two in the evening, every day. It took me fifteen minutes to get it into him. About thirty minutes later he would throw it up. It was a very slimy and heavy mucous. He kept on vomiting. We did this all outside. Then I realized; he was throwing up the toxins! After two weeks, my cousin came over with his two dogs and Spirgis started running around like crazy, dragging his hind legs all around. What a sight. This dog had a slipped disk, he had a tumor in his neck and a pinched vertebrae. None of the doctors had any hope. I mean this is a miracle dog!

I continued this for five weeks: wheatgrass in the morning and wheatgrass in the evening. Remember he was dragging his hind legs. The muscles were atrophied. We had to tie socks on them to protect them. Then, after the fifth week of giving him wheatgrass morning and evening, Spirgis got up, lifted his hind leg and started to walk. Eventually, he came with me on my [jogging] runs. He was healed completely. The vet could not believe this dog was still alive.

But that's just one story. In 1994, I got lyme disease. I actually got it in April but we didn't realize what it was until I was diagnosed in August. I felt like I had a huge hole in the pit of my stomach. I mean it didn't matter if I ate 10 lbs of food, I couldn't fill it. Later I learned that it's an emotional need that cannot be filled by food. But one night I stuffed myself with corn chips and apples and fruit and candy. Then I went to the Olive Garden, ate two salads and a plate of pasta. I was desperately trying to fill the hole in my stomach. I had to drive two hours so I drank Coca-Cola to keep awake. When I got home about 10 at night I started to throw up. I threw up 15 times, continuously. I was white as a sheet. I was completely off balance; the room started to spin. I was totally out of electrolytes. I told my friend: make me some water and juice a tray of wheatgrass. I put 8 ounces of wheatgrass into a 1,000ml enema bag of filtered water. And I did it again and again. Three times; it took me an hour and a half. I put it in and held 2,750ml total of water and 3 trays of grass in my colon and by the time I put the last tray in, I could see the color coming back into my body. It started at my feet and you could literally see it rising up. In any other circumstance I would have been bedridden for days. But 18 hours later I was out keeping my appointments and I felt absolutely alive.

When I feel tension and nervousness and emotional upheaval that I can't deal with, I take a wheatgrass enema and I come out a brand new person. I really feel I'm reversing the whole aging process. If I had not healed the emotional stuff, I'm sure it would have gone into a tumor or cancer. Wheatgrass is my miracle drug.

My husband takes Green Magma [barley grass juice powder] regularly. I don't. I only take it fresh; I like my home grown. I don't drink it at all. I just do my enemas. I could be working an entire day and be exhausted. I come home and I find my wheatgrass healing and cleansing and energizing. It's a powerful, powerful food.

Cancer of the Throat, Alimentary Canal and Melanoma

—Ruth Williams of Smyrna, Georgia

Let me relate a long, complicated and painful experience in just a few words. In 1961, I had a cancerous mole excised from my toe. Like the tree with its trunk being removed, the root system evidently remained. By 1976 my body was saturated with cancer. The melanoma had metastasized. It was in my throat and my alimentary canal. I could barely talk and it was impossible for me to take solid food.

[Mr. Eugene Williams] She was comatose and whittled down to only 62 lbs. We were in and out of five hospitals. She had chemotherapy and radiation. She was in an isolation room. Before I would go into to see her I had to shower and put on special clothing. Every vital life sign was gone. There was no possibility of living according to the medical doctors. They told me: Be prepared. It could be a very short time.

[RW] I did not want to stay in the hospital. I dropped down and prayed that the Lord would take me home. They sent me home to die with terminal cancer.

A neighbor of mine said: You don't have to die, and gave me a book called *How I Conquered Cancer Naturally* by Eydie Mae. With the aid of my nurse, we began growing and juicing wheatgrass and I started taking multiple wheatgrass enemas and implants daily. To keep up with this I had to grow 10 trays of wheatgrass daily. After a few weeks on this schedule, I had my first major healing crisis. I vomited continuously and it was awful, but I had faith in God.

Slowly, but perceptibly, I began to notice improvement. I was able to take small amounts of nourishment by mouth. Gradually, I increased the wheatgrass to 10 ounces per implant. Just six months after I started the wheatgrass program, I was able to walk again.

My situation was still severe and recovery was very slow over a long period of time. I got help from some very special doctors including Ann Wigmore whom I visited at her institute in Boston three times. As the raw food intake increased, I was able to reduce the frequency of the implants. In the years that followed I returned to normal weight.

[EW] It's 22 years later now, Ruth will soon be 75. She still uses wheatgrass and you should see her...in spite of where she has been, she is a picture of health.

[RW] I am alive and well. The Lord Jesus Christ spared my life. I am grateful to God for showing me the healing power of wheatgrass.

Lymphatic Cancer

—by Dennis Lampron, Chicago, Illinois

I was in school taking an airline computer course when I began to feel very tired and draggy. I began to lose weight, so I went to the doctor who did a lot of tests and couldn't find what was wrong. I noticed my lymph nodes were getting bigger and bigger, so they did more testing and found that I had lymphatic cancer which was spreading throughout my body.

The doctors began treating me with certain drugs which didn't work because I was sensitive to them. The cancer was getting worse and worse and they put me on chemotherapy. At this point, they gave me three months to live if I took the therapy. All my hair fell out, and I got big blisters because my immune system was destroyed.

Right about that time a lady told me about wheatgrass, which I had never heard of. I had never heard of live foods and I never ate salads. I was a 'McDonalds' person. She started bringing me wheatgrass and I began drinking the juice.

She also taught me to meditate to help with the pain. The doctors had me on morphine, but with meditation I stopped taking pain medication. I also realized that the lymph nodes were getting smaller. At that point I was directed to do a fast with wheatgrass juice.

I was very nauseous and had diarrhea. I was really cleansing. Then I went up to 2 ounces in 4 separate doses a day with distilled water in between and nothing else. I did that for thirty days and within that time my lymphatic system became completely clean. It took a bit longer to become normal, but the size of the nodes became normal. After that I added a variety of sprouts including sunflower and buckwheat and learned more about the diet from the Hippocrates books.[2]

The implants were interesting because that was where I got the pain relief. If I was in a lot of pain, I would get up and do an implant and within 45 minutes to an hour the pain was bearable. The wheatgrass juice implants also made it easier for me to breathe because the chemotherapy generated a lot of mucus in my lungs. It took about 3 months to get rid of it.

My thoughts were clearer. I was coughing up lots of mucus. In fact, this near-death experience actually introduced me to a whole spiritual world that I had no idea existed. I was enriched as a person. It took six months before the illness left completely. Six months ago my doctor tested me and could find nothing wrong.

It runs in my family to become gray early, but my gray hair disappeared. I could hardly believe it. My mother thought I was plucking the gray out. Not. At age 31, I look and feel younger than I ever have before.

> *I'm not going to no doctor. Last time I went to the doctor, he gave me so much medicine, I was sick long after I got well.*
>
> —Chico Marx, 'Horse Feathers'

Prostate Cancer

—Bill Nasdy, Bonita Springs, Florida

I had prostate cancer; my PSA was 14; the lowest it's ever been was 5. At first, I agreed to the surgery—major surgery to remove the whole prostate. This is February, the operation was scheduled for May. I also agreed to a program of intensive chemo and radiation. I'm 70 [years old] but I'm fit. First they put me on Flutamide. You know what that is? I found out later it's a chemical castrator. It was awful—the worst stuff. It sucked all the energy right out of me.

I went to a support group and somebody told me about wheatgrass. I went out and bought *The Wheatgrass Book* by Wigmore. Right then, I decided to stop the treatment and went full speed into natural therapy. I canceled the surgery and got off the drug after 5 weeks of that hell. I changed everything. They told me: you're going against doctor's orders. Well, I've become a vegetarian now which is a big deal for me because I grew up in meat country. Right away my PSA went down. (I test it every month.) I'm taking 6–8 ounces a day [wheatgrass]. I take it before every meal. It's changed my whole life. I'm also taking ozone therapy at the Optimum Institute in Naples. It's fantastic and they don't even charge for it. And I'm pouring on the antioxidants, too—Essiac tea, CQ10, shark liver oil, vitamin E, saw palmetto and 9,000mg of vitamin C daily. Antioxidants is the way to go and wheatgrass is of course loaded with them.

You know I forgot to tell you, four years ago I had microwave treatment done on my prostate to relieve painful urination. It was not successful at all and I strongly believe it had a lot to do with creating the cancer. It wasn't approved by the FDA then but it's accepted both here and in Canada now. I would never do that again and I definitely don't recommend it. This is the trouble with the medical establishment. They let you take damaging treatments even before they're officially approved and then approve them, while nontoxic wheatgrass and ozone are outlawed. We've got to take back control of our health from the government and the doctors. They're not gods. It's unbelievable that the public trusts them the way they do. They're pushing dangerous drugs and dangerous therapies and unnecessary surgery. I mean c'mon, this is nuts! But the tide is turning. The middle class is getting on the bandwagon. Look at me, I never knew about this stuff!

I'm in remission now. On my last PSA test which is the newest type of PSA...the nurse says to me: I hope you didn't do any exercise today. I had just come from doing 50 laps in the pool! Turns out exercise artificially elevates the PSA. I said hell, $58 bucks down the drain. Guess what? I got a call the next day, my PSA was 0.12.

Listen, I'm an active guy for my age, but I was dragging. You wanna know what the real surprise for me was? Since the wheatgrass I've got twice as much energy. I sky dive; I work a full time job; I swim 50 laps. And let me tell you, it's all about exercise and diet. Your immune system will take care of you if you take care of it. That's the key. And you can't do it if you're bombarding yourself with smoking and drinking and bad food and high stress and toxic drugs. Chemotherapy kills the cancer but as soon as you stop, bang–it comes right back somewhere else and it's worse.

I got 8 trays of wheatgrass growing right now and I wanna tell you something else. Since I've started this, I've never had a cold, never had the flu. My wife was sick but I never caught it. Cuts and sores heal up immediately; it has definitely changed my life. I also take a carrot juice every day with garlic—lots of garlic—beets, parsley and sprouts, lots of sprouts, in the salad especially. No dairy, no beef, no pork, no chicken—the only thing I'll take is a deep sea fish occasionally. I'm a believer in it [vegetarianism]. Give me the grains and vegetables; give me the grass instead of waiting for the cow to eat it and then eating the cow.

Lupus, Hysterectomy

—Cynthia Gebhart, Geneva, Florida

I have lupus—I don't know if you know what that is? It's when the antibodies in your blood start attacking your normal cells. The symptoms are everywhere. I get rashes on my skin, my joints swell up, headaches, shortness of breath, real flu-like symptoms, depression...I just feel lethargic all the time. I'm a real active person, too. Ask anyone, I'm always in motion. I got a full time job. I help my husband out; I got two kids; we live on the St. Johns River. But this lupus just lays me flat on my back and anything can trigger a bout of it. I'm walking on egg shells most of the time.

A year ago, after I had a root canal, the lupus flared up and wiped me out for months. Two years before that, I had a cyst removed on the right ovary. It took me five months to recover from that. I had such severe pain; I could not get beyond the pain and nothing I did worked. I was tired and run down all the time. My joints swelled, my skin tissue got inflamed. I had the desire to work but not the stamina. So, I was definitely afraid of a hysterectomy. I watched my aunt and my sister both go through it. I sat with them while they moaned and groaned for six months. And they didn't have lupus! So I was paranoid as hell about this surgery.

So I figured before I go under the knife, I gotta try to control the lupus. My husband and I got on the internet and we looked up all the lupus doctors and we picked out the best one. I mean this guy was world renowned, head of his department, a prestigious hospital....the best. So I dragged myself out of bed and we went to Atlanta, Georgia. The doctor examined me and said: You're dying. I want you to go straight to the hospital. When I arrived, they started pumping me full of steroids and two different types of drugs and IV's. Then I *really was* dying! When I got out of that hospital I slept for two weeks. I could not get out of bed and I'm a perpetually active person. My husband kept saying: You were not this sick before you started all these drugs. So we went back to Atlanta and he checked me again and put me back in the hospital again. I kept on asking him why? And I never felt like I got a good enough answer. I kept trying to tell my husband: He never gives me a straight answer. That's not right. He just kept pumping me with more drugs. I was looking and feeling worse. This was another two weeks in the hospital. My husband says: You were never this ill before; we need to get you off these drugs. By the grace of God, that's when we ran into Russell.[3]

Russell had put my uncle on wheatgrass so I dropped the drugs and decided to gave it a try. I used to tease my uncle; I'd moo like a cow. But I wanna tell you, three weeks into drinking the grass juice and I felt like I could climb mount Everest. I mean, you have no idea how long it's been since I've felt this good. I felt like I could touch the clouds. My worst fear was that it was a fluke and it would end. Then I started echinacea and ozone. My husband and I were totally flabbergasted by all the wonderful information and therapies we were discovering. And my GP [general MD] could not believe my cholesterol. I'm always testing it because I have this thyroid disorder. It's always been between 240 and 280 for years. All of a sudden it's 173. After the wheatgrass—that's the only thing I did different! Of course, I got off the drugs. He told me, by the way, that I was taking enough antibiotics to kill an elephant. So much for big shot specialists! Now, I take nothing unless it's natural. I'm growing the grass and juicing it and celery and carrots and taking different herbs and I feel great. Yes, I did go through with the surgery [hysterectomy] and this is the most unbelievable part. Not *only* did I *not* have a lupus flare-up, but in two weeks after that major operation I was back home stripping my bed and cooking food and I haven't stopped since. There is absolutely *no* doubt in my mind that wheatgrass was the thing that turned everything around. I've been through too many operations, hospitals and too much hell for anyone to tell me otherwise.

—

The effect these highly nutritious green drinks are having on all my patients, especially my arthritis patients, is nothing short of amazing... I tell you, no matter what your age or present condition, these grass superfoods can quickly take you to a whole new level of radiant health—sparkling eyes, abundant energy, pain-free joints and a zest for living you remember from the healthiest days of your life. —Dr. Julian Whitaker, MD, editor of *Health & Healing Newsletter*.[4]

—

As measured by present standards, our diet is better than ever before in history. Yet, we are fighting a losing battle against degenerative diseases. Cancer and heart disease have worsened with every decade and this is in spite of all our knowledge about vitamins and advances in medical science. There is something missing in our diet and it may well be the grass leaf factors.—Dr. Charles F. Schnabel, 1939.[5]

—

The art of healing comes from nature, not from the physician. Therefore the physician must start from nature with an open mind. —Paracelsus, 1493–1541[6]

Healing Resorts

Wheatgrass and Buckwheat greens at the Hippocrates Health Institute, in sunny West Palm Beach, Florida

Sanctuaries for recuperation from chronic disease and restoration of health have been around for centuries. Hippocrates, the father of modern medicine, is quoted as saying: Rest is sometimes the best remedy. The Chinese have an ancient adage: Healing is three parts treatment, seven parts nursing. In modern times, American physician Edward Trudeau became famous for curing tuberculosis at his sanitarium in the fresh air of New York's Adirondack mountains. This was 60 years before the development of the modern drug treatment. John Harvey Kellogg, co-founder of the corn flakes company, was a prestigious surgeon from Bellevue Hospital Medical College in New York. But he had a passion for nutrition, vegetarianism and natural hygiene and ran two popular sanitariums in Battle Creek and Miami. The treatments in these places were fresh air, rest, exercise and healthy diet. When focus on the body, mind and spirit is intensified along with insulation from the stresses of society, miracles happen. Our society places great emphasis on outside appearances—skin, cosmetics, muscles. If the color of our lungs, the hardness of our livers or the congestion of our colons were visible to all, we would place greater emphasis on keeping our insides in pristine condition. Every middle class family in America is stretched to their limits paying health insurance bills. But the best insurance of all is a couple of weeks a year at a health resort where you can cleanse, nourish, rejuvenate and heal. The following are holistic lifestyle resorts where wheatgrass is a major part of the program.

Hippocrates Health Institute

561-471-8876, fax 561-471-9464. 800-842-2125 (reserv.), 1443 Palmdale Court, West Palm Beach, FL33411. http://www.hippocratesinst.com/

After my stay at Hippocrates, I continued following the program strictly. Not only did I eventually get rid of my malignant breast tumor, I also increased my energy level, improved my skin tone and hair volume. Most importantly, I have been cancer-free for nearly a decade now. I return to the Institute twice a year for reinforcement and reaffirmation.

—Rachel Budnick, Chicago, Illinois

Nestled in the quiet woods of West Palm Beach Florida is an oasis of health and healing where people from all over the globe convene to mend their bodies and revive their selves. This serene 30 acre sanctuary carries forth and expands upon the visions of Ann Wigmore and Viktoras Kulvinskas who founded the original Hippocrates Health Institute in Boston in the 1960's. Brian Clement, its director, was one of Dr. Ann's finest managers, but took the bold step of leaving her and striking out on his own 1,000 miles south in the sunshine state. Today, the 'new' Hippocrates is a powerful beacon lighting the way for thousands to find healing through living foods and wheatgrass.

A visit to Hippocrates typically involves a three week 'Life–Change Program' of healing without drugs. Even though there are two medical doctors on staff (bring your medical records and drugs), this is not a medical facility. It is closer to a spa than a hospital. You must be committed to working on yourself and be self-sufficient (or come with a personal assistant). They view themselves more as mentors than doctors. While hospitals may sicken you with their food and medicines and depress you with their manner and ambiance, this place makes you come alive with its exquisite natural surroundings, positively motivated good company and therapies that are enjoyable rather than painful. They have three pools purified with ozone instead of chlorine, a whirlpool, sauna and a 'vapor cave' inspired by native American medicine men. Their exercise room has treadmills, rebounders, a vibrosaun, a hydrosonic bed, bio-rhythmic equalizer and you can always jog on the nature trail.

Their therapy building is located by a lake and includes colonic irrigation, massage, hydrotherapy, acupuncture, reflexology, homeopathy, chiropractic, kinesiology, psychotherapy, darkfield microscopy, yoga, shiatsu, deep-tissue massage, lymphatic drainage and polarity energy balancing. They also offer magnetic therapy in which you lie inside a magnetic field that increases the ion exchange between the inner and outer cell walls enhancing cellular function even in hard to reach places. While healing to the body, these therapies also relax the mind and enhance the spirit.

Hippocrates grows its wheatgrass in an air-conditioned greenhouse with grow lights. Their wheatgrass bar is open 24 hours per day and is self-service. Although it depends on your needs, expect to go on a juice fasting, detoxification program with colonics, enemas and wheatgrass implants. When you're ready to eat solid food, it will all be life-giving foods. Expect to be delighted. If raw foods sound boring to you, your three weeks at Hippocrates will teach you how to prepare feasts that will nourish your cells in addition to pleasing your palate.

Hippocrates Buffet—Neither Cooked Nor Boring

Almond Basil Loaf with Red Pepper Coulis
Homey Hummus
Stuffed Avocado Platters
Cauliflower and Mushrooms a la Greque
Rose Sauerkraut
Sunflower Nori Sushi
Spelt Tortilla Rolls
Sliced Red Bell Peppers
Dulse (Purple Sea Vegetable)
Gorgeous Green Salad
Sprout Medley
Fresh Corn on the Cob

If you are on a juice diet, you will dine on the finest organic veggies and sprouts available, plus watermelon juice, organic herbal teas, lemon water and, of course, wheatgrass. There are several food preparation lectures, demos and sampling throughout the three-week course.

Accommodations are some of the most exquisite available for a wheatgrass retreat. They have a wide range of apartments and guest houses both on and off premises. All have a stucco, terre-cotta theme. Many of the off premise facilities are within walking distance. Some include private baths, even sunken tubs and jacuzzies. Even the most economical rooms are perfectly comfortable.

Hippocrates also offers a Health Educator's Program which is a nine week training for those who want to incorporate some of the lifestyle, nutritional and therapeutic disciplines practiced at Hippocrates in their own profession. A fair number of people actually change careers and start growing wheatgrass and sprouts or become nutritional counselors or pursue other alternative health professions.

After your stay or even before, you can keep up with the Hippocrates program by becoming a member. Membership fees are very modest and you get mailings, including their newsletter. To give you a sense of it, here are some topics featured in recent newsletters: Light–Medicine of the Future; What We Know Now About Food That We Didn't Know Then; Staying Flexible in Body and Mind; Relationship-Voyages Through Life; Self-Esteem-The Mirror of the Mind.

Occasionally, Hippocrates will close to the general public and devote a life-change program strictly for AID's patients. According to director Brian Clement, several are now leading fully functional lives.

It's not only the nutrients and phytochemicals in wheatgrass that make it therapeutically viable, it's the oxygen and electrical properties of the fresh living juice. —Brian Clement, director.

Optimum Health Institute - San Diego

6970 Central Avenue, Lemon Grove, CA 91945-2198. 619-464-3346, Fax 619-589-4098, Reservations 800-993-4325. http://Optimumhealth.org

Before this way of living began for me, I was well over 200 lbs., and survived on any pill, fast food, alcohol, or any other relief available. I was a captive in my own house and in my own body. I had five surgeries, all related to food and self-abuse. Today, I am a trim, vibrant, fifty year old woman, in love with life. I have friends who, including myself, would not be alive today if it were not for the Optimum Health Institute.—Katie

Next to Ann Wigmore's original institute in Boston, this is the oldest wheatgrass retreat center in the world. After her stay with Dr. Ann in 1976, Raychel Solomon was inspired to establish a similar center at the opposite end of the country. Raychel joined with Robert and Pamela Nees to form what was initially called the Hippocrates Health In stitute of San Diego and is now known as the Optimum Health Institute (OHI). Fifty thousand people have passed through their doors since then, and all of them have experienced the benefits of wheatgrass. A maximum of 200 guests can stay here at one time making it the largest wheatgrass retreat center in the world. Not only that but it's full most of the year and they all come by word of mouth. There is a feeling of community here, alumni often come back for 'tune-ups.' The staff is so loyal, some came as guests twenty years ago and never left. Technically, the institute is a not-for-profit, nondenominational church and as such, everything here is very affordable. They have been pioneers in making wheatgrass juicers and are the only ones with a 100% stainless steel machine.

OHI's master grower Michael Bergonzi in the 1,500 tray Wheatgrass Cathedral

Being relatively close to Hollywood, this center has had many celebrities walk through its doors. Ben Vereen, the Beach Boys, Mrs. Sammy Davis, Jr., Edie Adams (wife of Ernie Kovacs), Lane Smith, Debbie Reynolds, Joel Douglas (son of Kirk) and Ernie Banks of the Chicago Cubs are just some who discovered wheatgrass at OHI.

The philosophy and methods of this institute are very close to those of Dr. Ann. They believe that, given the chance, the human body is self-regenerating and self-cleansing and that to improve the physical body, you must also work on the mental and spiritual. Balance and harmony are the goals and detoxification, wheatgrass and living foods are the means.

Classes

Although you can stay more or less, the basic recommended program here is three weeks. It takes that long for the body to begin the detoxification process and for the person to learn the entire program well enough to take it home. The first week's classes are primarily theoretical in nature—how the body works, how the mind works. The second week, classes are more practical, such as how to make rejuvelac and how to grow wheatgrass. During the third week, guests get actual hands-on experience in making raw food recipes so they can continue when they go home. A fourth week, offered at OHI's Impero beach location 20 minutes away, focuses on integrating the lifestyle changes back into society.

Three Weeks to Rejuvenation

First Week	Second Week	Third Week
Daily Exercise	Daily Exercise	Hands-on food preparation
Mental Detoxification I	Sprouting Instruction	Sauerkraut making
Elimination	Your Life as a Gift	Dehydrating Foods
Testimony	Mental Detoxification II	Seed Sauces
Emotional Detoxification	Fermented Foods	Mind-body connection
Digestion	Emotional Detoxification II	Transition Diet
Instant Relaxation	Wheatgrass Planting	Equipment
Pain Control	Communication	Self-Esteem
Know Your Body	Organic Gardening	At-Home Follow-Up
Food Combining	Personal Care	
Wheatgrass Instruction	Menu Planning	

Therapies and Food

This institute sticks to the basics. OHI operates on the principle that the body heals itself if given the proper tools. Their program of detoxification and raw foods is designed for self-healing and re-balancing. Wheatgrass juices, enemas and implants are the essential therapy here. They believe that the grass detoxifies and nourishes and that is all the body needs to normalize health. It's the oxygen, the nutrition and the life force in grass that make it so potent.

They do have massage therapists, chiropractors and colonic hydrotherapy on the premises. These services are run as independent businesses and are paid for separately. They have a jacuzzi and a skin care center for body wraps and facials to detoxify the largest organ of elimination, the skin. The exercise room is low impact and focuses mostly on cleansing the lymph system via rebounding on a trampoline-like device.

The diet follows strict food combining principles and is all raw. Fermented foods such as rejuvelac, sauerkraut and seed sauces play an important role in the diet aiding digestion, elimination and colon health.

Buckwheat and sunflower sprouts, greens, fruits, and fresh vegetables are available seven days per week. A juice fasting liquid diet is recommended during the first few days.

The Wheatgrass Cathedral

The stature this greenhouse holds at OHI is reflective of the importance of grass to their healing program. It is a state of the art building designed by master grower Michael Bergonzi. This temple of green, houses 1,500 trays and harvests 150 new ones per day. The sun is its only light source and the temperature is controlled by shades and circulating air. The science of growing wheatgrass indoors has been refined down to the most minute details, from seed quality to soil quality to the speakers in the four corners that serenade the grass with classical music.

Accommodations

The accommodations here are simple and functional minus the finery and trappings of expensive vacation retreats. OHI focuses on the inner-life space, not the outer one. Double and single occupancy are available and prices are most economical. They have suites and town houses in addition to apartment dwellings.

Medicines

This is not a medical facility. There are no doctors on staff. OHI does not treat disease. Instead they will help you marshal the physical, mental, emotional and spiritual resources that enable natural healing. You must be able to take care of yourself, attend classes and be open minded to learning new ways of living and eating. Patients undergoing chemotherapy should not attend. Whereas chemotherapy compromises the immune system, wheatgrass and the other therapies used at OHI strengthen it. You can do the program before chemotherapy to strengthen your system or after to rid your body of chemotherapy's toxic effects. As far as continuing with your prescription drugs, you must make your own decisions. Nutritional supplements are not allowed because they impose upon the body's natural chemistry. Wheatgrass, on the other hand adapts to the body's needs in addition to providing a broad spectrum of over 80 different nutrients. This program is a truly holistic one, addressing all aspects of a person's health—mental, emotional and physical.

Not only is this center the largest of its kind in terms of capacity, but they have expanded into two other locations as well. They have 30 apartments on beautiful Impero Beach only 20 minutes away for graduates of the basic program and have a brand new facility in Austin Texas.

> *We don't address the illness per se. The human body is the greatest doctor on earth. We remove the roadblocks and the body starts healing itself. Then, you name it—aids, cancer, arthritis, there are no limits.*—Robert Nees, director OHI

Optimum Health Institute - Austin

512-303-4817, Fax 512-332-0106. http://Optimumhealth.org
Route 1, Box 339-J, Cedar Lane, Cedar Creek, TX 78612.

As you place your feet on the massive stone floor entry, you begin to experience the quality and grandeur of the Optimum Health Institute in Austin, Texas. Exquisite, yet rustic elegance is evident everywhere, from the custom built furniture, to the atrium court and the oak stairway. The whole place is a labor of love. Cory Carson, son of the entertainer Johnny Carson constructed the center as a memorial to his brother Rick. His generosity helped make it possible for OHI to obtain the campus.

Nestled in a park-like setting on 14 beautiful acres, the center is less than 20 minutes from Austin's new airport, offering both convenient access and country charm. Despite its proximity to a metropolitan area, there is a sense of peace and privacy here. A spiritual atmosphere is evident as you drive onto the beautiful grounds through a winding entry that reveals a large exquisite stucco and tile building. Its mission style architecture, textures and earthy colors add to the overall feeling of relaxed opulence.

This place was designed and built to be a health center. There is an exercise room with state-of-the-art equipment that overlooks the swimming pool and spa. The 20,000 square foot mansion houses 30 beautifully furnished bedrooms with all the amenities. It's obvious that quality standards for furniture, fixtures and equipment have not been compromised. There are also two distinctive suites that are just beyond a graceful and inviting walking path.

Austin OHI is managed by the same family as San Diego OHI and is also a not-for-profit facility. It has the same philosophy and program as San Diego and tuition is comparable. Both emphasize the holistic approach of physical, mental and emotional health, along with detoxification and wheatgrass. But the atmosphere in Texas is definitely more elegant and intimate.

Naples Institute for Optimum Health & Healing

800-243-1148. 941-649-7551, fax 941-262-4684. 2329 Ninth Street North, Naples, FL 34103 USA. http://www.naplesinstitute.com

With its grand opening in May of 1998, this is one of the newest retreat centers that offers wheatgrass therapy. Located on the pristine, sunny gulf coast of Florida, it is a natural environment for healing. It has the unique contrast of being in an active vacation area, yet hidden in seclusion by its location and landscaping. The Garden of Eden must have looked like this: Tropical flowers, lush greenery, crystal water fountains, sculptures scattered throughout the grounds, and cool breezes from the Gulf of Mexico adding a tinge of salt to the air.

The institute was founded by Russell Rosen, a long time advocate of natural healing and frequent guest of the Optimum Health Institute of San Diego. Like OHI, this institute (NI) provides a program of cleansing, living foods, exercise, education and meditation for full physical, mental and spiritual healing. Like the other centers, NI is not a medical facility. There are no medical doctors on staff, but numerous other health professionals provide custom care.

The crystal clear ozonated pool at Naples Institute in Florida

This institute is unique in that you are not required to check in and out on specific days. Come any day you wish and stay as long as you need. But remember, you didn't get sick overnight and it will take time to mend. The program here lasts from one to four weeks and is loaded with all kinds of fascinating therapies. You'll be thrilled and pampered by the amount of personal attention you receive. First, you'll get a live blood cell analysis, then a massage, a wheatgrass juice, a session with a psychologist and a homeopath, not to mention, yoga, Qi Gong, guided meditation and a walk along the crystalline beach.

Diagnosis

Live Blood Cell Analysis and Biological Terrain Assessment are two of the main diagnostic tools, here. You can actually see your blood cells moving on a computer monitor. Skilled herbalist Charles Marble, studies your results for circulation problems, nutrient deficiencies, protein digestion, viral and fungal activity, toxicity, liver problems and acidity-/alkalinity. Charles believes that hyper-acidity at the cellular level is responsible for most disease. Otto Warburg (1931 Nobel prize) discovered that cancer cells are highly anerobic and thrive in an acidic environment. Acidic cells lose their potassium, magnesium and sodium and fail to efficiently convert oxygen into ATP for energy. These diagnostic tests for which you have to fast for at least 12 hours before provide a biochemical map of cellular health. Armed with this incisive but non-invasive diagnosis, you begin the process of education that leads to correction through a variety of therapies.

Therapies

People have enjoyed the benefits of therapeutic massage for thousands of years. But most associate massage only with relaxation. We have forgotten that massage stimulates circulation, eliminates toxins and raises immunity. Listen to this smorgasbord of hands-on body therapies.

Body Talk is a treatment that views the body as a complex series of circuits and energy systems. When the circuits of communication between systems malfunction, disease, lower vitality, and long-term breakdowns occur. Body Talk reestablishes the lines of communication (nerves, meridians, electromagnetic waves, etc.) through a synthesis of yoga, acupuncture and applied kinesiology. It balances you.

Myofascial massage is a gentle hands-on technique that provides sustained pressure to eliminate pain and restore motion. The fascia is a web of fibrous connective tissue that spreads throughout the body. It has the potential to influence every muscle, bone, nerve, blood vessel and organ.

Manual Lymph Drainage Massage (MLD) redirects stagnant lymph fluid and increases the transport capacity of the lymph. The lymph system is crucial to having a healthy immune system and to cleansing the tissues.

NI also offers Qi Gong (Chee Gong), a breathing and body movement exercise from ancient China. Regular practice influences immunity, stamina and mental clarity. There is also morning meditation and yoga including aqua-yoga in the pool.

Swimming is allowed under this miniature bridge at Naples Inst.

Ozone Therapy

This is one of the things that makes this institute special. Ozone is an aggressive form of oxygen that occurs in abundance after thunderstorms. Although its use in the USA is limited, it is quite prevalent in Canada, Germany and Italy. Some of Europe's most prestigious healing centers have been using therapeutic ozone for a long time. There are thousands of documented cases which show that ozone reverses illness caused by viruses, bacteria, parasites, autoimmune deficiencies and even degenera tive diseases such as cancer and AIDS. In essence, this therapy increases oxygen levels in the blood and at the cells. This is also one of the reasons wheatgrass is effective. Many diseases have their root cause in cellular starvation for oxygen and the resulting accumulation of toxins. While therapeutic ozone is not prevalent in the USA, ozone air and water purifiers are available and ozone is an approved alternative to chlorine as a bacteriostatic for pools.

Wheatgrass Bar & Classes

Imagine soaking up the sun lounging by the pool sipping a wheatgrass cocktail. The juice bar here, which opens at 8 a.m. is dispensed from a palm thatched hut at poolside. Its menu includes organic raw salads and any juice combination you can think of. But don't fall asleep, you'll miss the lecture. Classes are held throughout the day and evening on topics like wheatgrass growing and juicing, colon health, enemas and implants, live-food preparation, herbal alternatives to prescription drugs, self-esteem, etc.

Accommodations

This institute is on the property of the Naples Best Western Inn & Suites. There are thirty condo apartments available and thirty deluxe suites with mini kitchens. Prices here are reasonable standard hotel prices and a lot of amenities in the program are included in the package. You also have the flexibility to order a la carte. This is a place where the family can easily entertain themselves in the outside vacation playground of Naples while you do your inner work within.

We provide an environment where people can be educated to care for their own health. They take home the skills they learn and incorporate them into their new lifestyle. We are empowering them to heal themselves and to prevent their own future health problems. —Russell Rosen

Ann Wigmore Institute – Puerto Rico

PO Box 429, Rincon, Puerto Rico 00677. USA. 787-868-6307
fax 787-868-2430. E-mail: wigmore@caribe.net

How can I go back to the way I was eating? I'm feeling so light and energized. This lifestyle is making more sense to me now. The staff here is so devoted... I can only imagine what I'm going to do with all this new energy I'm discovering inside!

Imagine waking up to the sound of waves crashing on the beach and beginning your day with yoga by the sea. This is a refuge of beauty and tranquility where there is a natural propensity for inner purity so as to achieve an equilibrium with the pristine surroundings. This haven for purification of body, mind and spirit was one of the many gifts Dr. Ann Wigmore left us. Today, her spirit is still palpable in the cool breeze and her teachings are carried on faithfully by her devoted followers.

Here you will learn the basics of the Ann Wigmore self-healing and living foods lifestyle program in its original, unmanipulated form. The basic program is two weeks, but one week intensives and one month discounts are also offered. You can also just go for a week of personal healing which is less intensive. This institute, the Ann Wigmore Foundation in New Mexico and the Hippocrates Health Centre in Australia, teach Dr. Ann's Living Food Lifestyle program in its original form. Much of the teaching information described here applies to all three of these institutes. The premise is that natural, uncooked, and unprocessed food provides all of the living enzymes, vitamins, minerals, proteins and amino acids necessary to nourish, protect, and defend the body's internal and external well-being. When the body is cleansed and nourished, the symptoms of disease disappear. Students learn about eating and cleansing and how to become self-sufficient in growing and maintaining a lifestyle that encourages positive, physical, emotional and spiritual well being.

Photo by David Kelman

Rejuvenation comes not only from living foods at the Ann Wigmore Institute in Puerto Rico, but also the ocean.

Accommodations here range from private with full bath to semi-private and dormitory. Prices are by the month or the week and include the program, transportation to and from Mayaguez or Aguadilla airports and meals. There are three simple, nurturing living foods meals each day that include sprouted organic sunflower and buckwheat greens, alfalfa and other grain and legume sprouts, organic garden grown veggies, fermented nut and seed yoghurts, veggie loafs, snacks and foods prepared in the dehydrator rather than the oven, energy soup and rejuvelac.

Rejuvelac, energy soup and wheatgrass are the cornerstone foods of Dr. Ann's Living Foods diet. Rejuvelac is a mildly fermented drink made from sprouted wheat before it reaches the grass stage. It is the fine wine of living foods because it is aged into a rich brew of active enzymes and friendly bacteria. These pro-biotics are more effective than anti-biotics in fighting disease. Instead of killing bacterial invaders with anti-biotics,

The Ann Wigmore Living Foods Curriculum

Live food preparation	Dehydrating foods	Yoga and gentle exercise
Wheatgrass growing	Energy soup	Breathing and relaxation
Indoor Gardening	Composting	Lymphatic exercises
Sprouting	Iridology and reflexology	Skin brushing and care
Enzyme nutrition	Detoxification	Lifestyle adjustment
Digestive health	Internal cleansing	Positive thinking
Rejuvelac	Colon health and hygiene	Meditation & visualization
Fermented foods	Enemas and Implants	

pro-biotics compete against them until their numbers diminish and their threat is quashed. If there are more cops than burglars, the neighborhood is safe. Here you will find lactobacillus, acidophilus, bifidus and many other friendly bacteria. They help control the intestinal environment against invading microbes, yeast, fungi, parasites and also synthesize important B-vitamins. It's germ warfare in there and rejuvelac is your best armament. Yes, this is the same kind of protection you get from yoghurt but rejuvelac is easier to digest, more concentrated and more economical in the volumes required to change your inner terrain.

Energy soup is a thick shake or cold soup of super nutritious foods that are also easy to digest. Every food requires energy to digest. The ratio of energy received vs. energy spent on digestion determines how much net energy you will profit. If you fall asleep after eating a big meal of meat and potatoes, you have spent more than you earned. No matter how nutritious the meat is, you've lost. If this was the stock market, you'd consider it a bad investment. Energy soup is like a golden parachute—your energetic profits are long lasting. The ingredients? It's a blended salad/soup of dulse, sprouts, indoor sunflower and buckwheat greens, organic garden greens, apple or papaya, avocado or almond cream and rejuvelac. There are many versions.

Dr. Ann considered wheatgrass the most important nutritional vegetation on the planet. Many attribute the maxim to her: Fifteen pounds of wheatgrass is equal to the nutrition of over 350 pounds of vegetables. (These are actually the words of Dr. Charles F. Schnabel.) Wheatgrass is an excellent source of crude chlorophyll; a good delivery agent of cellular oxygen; rich in digestible amino acids; abundant in over 80 nutrients and phytochemicals; a potent restorer of cellular and blood alkalinity; a powerful liver and colon detoxifier; and an electrically active, high vibration living food.

The Ann Wigmore Foundation.

PO Box 399, San Fidel, NM 87049. 505-552-0595, Fax 505-552-0595.
http://www.wigmore.org/~wigmore/ Email: wigmore@wigmore.org

As a director of the Ann Wigmore Foundation, Shu Chan is as close a descendant to Dr. Ann as anyone. They were very close and in the final few years of her life, Ann had the highest faith in Shu. Even after Shu had returned to Taiwan, Ann asked her to come back and run the foundation. Shu is Dr. Ann's chosen successor and judging from the feedback, she deserves that distinction.

Dr. Ann founded the Hippocrates Health Institute in Boston in 1969 and later changed its name to the Ann Wigmore Foundation. She also started the Ann Wigmore Institute in Puerto Rico. Since the Boston fire in 1994, the foundation has been closed (*see Epilogue*). But now, the Ann Wigmore Foundation has been reborn in the spiritual heartland of New Mexico, just one hour west of Albuquerque.

The main learning center is an adobe building with rooms for dining, kitchen, sauna, colonics, massage and a central common room. The other two buildings on the property are residences for guests and staff. These natural wood octagon structures provide for 8 private and semi-private rooms and dormitories. There is room for over 20 students with 6–7 staff so there is plenty of personal attention. Their greenhouse is a geodesic

Basic Health Principles of Ann Wigmore

Wheatgrass. Taken daily in therapeutic dosages of approximately 8 oz daily either orally or by rectal implant. Dosage will vary according to the condition.

Detoxification. Colon cleansing through colonics, wheatgrass and coffee enemas, wheatgrass implants, liquid diet and raw foods.

Living Foods. Cooking and processing kills living enzymes. Raw, uncooked, unprocessed. Whole grains not flours. Fresh vegetables not frozen or canned.

Exercise. Gentle exercise to improve blood and lymph circulation and help cleansing.

Juicing. Daily juicing of fresh vegetables, sprouts, greens and watermelon.

Sprouts & Indoor Gardening. Create your own year round indoor garden. Young vegetables and grains in their (nutrient) prime.

Body, Mind, Spirit. Body cannot cure without the mind and spirit. Dr. Ann was a doctor of divinity. Includes positive thinking and meditation.

Fermented Foods. Digestive health can be maintained by the influx and dominance of these friendly bacteria rich foods on the intestinal terrain.

Sunshine, Water & Oxygen. The energy source for all life. Without a healthful intake of these vital elements there would be no life.

dome structure which is heated in the winter by solar energy. The New Mexico climate is very temperate. Although it can be 95°F in the summer, the inside rooms are cool enough even without fans. Winters average 50–60°F most of the time. With the vegetable garden, the extended New Mexico growing season, solar energy and the indoor sprout gardens, the institute follows Dr. Ann's principles of self-sufficiency.

Here, the original, unrevised Ann Wigmore program can be found. The one to three week program emphasizes Dr. Ann's philosophy of restoring health through cleansing the body and re-nourishing it with living foods. Teaching is geared to a learn-by-doing style. Classes are about the growing, juicing, and preparing of easy-to-digest, live-enzyme foods along with colon cleansing including wheatgrass enemas and implants. You'll also learn sprouting, indoor gardening, nut and seed fermentation, blending, dehydrating and composting. The emphasis is on successfully getting the student to a place where you can continue the diet and lifestyle after your return to the larger society.

Mainstream eaters would term this food fare Spartan, but in fact it is very creative and definitely delicious. First thing everybody has is wheatgrass. Then there is a smoothie cereal, later energy soup, organic salad, sprouts of all kinds, seed, nut and veggie loafs, nori rolls, homemade sauerkraut. The sauerkraut is actually veggie-kraut made from different fermented root vegetables. The fermented foods, which include rejuvelac and seed sauces, provide an army of friendly bacteria that reclaim your digestive tract from putrefaction, giardia, parasites and yeasts. (Rejuvelac is made from young wheat sprouts before they turn into grass.) One day a week is just a liquid diet and even that is loaded with variety and taste. A liquid salad is just as delicious as a real salad only several times more concentrated in flavor and nutrition because the vegetables are all juiced. There are hands on recipe classes for making rejuvelac, seed sauces, veggie and nut loaf, energy soup, almond milk, almond cream, dairy-less ice cream, and flour-less celebration desserts. It's quite gourmet.

Exercise, yoga, meditation and massage are sprinkled throughout the program. As with most of the other institutes, there are no medical doctors here. Neither is there any diagnosis. There is only one treatment for all illness—cleansing and renourishing of the whole person—body, mind and spirit.

From the medical evidence and years of experience in observing people with all kinds of problems alleviate their symptoms using fresh chlorophyll from wheatgrass, I am convinced that young grasses, alfalfa and other chlorophyll-rich plants are a safe and effective alternative treatment for ailments such as high blood pressure, obesity, diabetes, gastritis, ulcers, pancreas and liver problems, osteomyelitis, asthma, eczema, hemorrhoids, skin problems, fatigue, anemia, halitosis, body odor, and constipation. I have found chlorophyll to be effective in alleviating symptoms of oral infection, bleeding gums, burns, athlete's foot, and cancer. —Dr. Ann Wigmore[1]

Hippocrates Health Centre of Australia

Elaine Ave, Mudgeeraba 4213, Gold Coast, Queensland, Australia.
Tel. (61)075-530-2860

I've lost a lot of weight in four weeks here. I've stopped smoking once and for all. My skin is clearer. My eyes are shiny again. Chronic fatigue is gone. I feel fully energized. I've gained an incredible amount of knowledge about my mind and body and I've developed the self-discipline necessary for me to continue to blossom at home. —Jennifer Blake, Hornsby, NSW

The Hippocrates Health Centre of Australia was started in 1985 with the blessings of Dr. Ann Wigmore. Ronald Bradley is a faithful adherent to the principles set forth by Dr Ann. As proof, Dr Ann gave him the use of the Hippocrates name and logo—something that she did not part with easily. Here you will find the unblemished version of Dr. Ann's teachings in all its simplicity and wisdom.

The recommended program here is three weeks. But there is pricing for one to six weeks. Experience has shown that the longer you stay the better the success in maintaining the diet/lifestyle after you return home. This institute closely follows the Ann Wigmore program. (For details, see the program described under her name in this chapter.) And what a bargain it is for Americans willing to fly over. A thousand Australian dollars requires only $600 American (in 1998). That shortens the psychological distance of this otherwise far down under retreat. Be prepared to write a letter or make a phone call. Public faxing, website and e-mail are not available. But there are some discount phone companies that offer low per minute rates to Australia to help lessen the expense of inquiries and arrangements.[2] Tuition fees include most everything you'll need including an iris analysis, private health assessment and personal counseling. Massage and other similar services are extra. Check your clock before you call. They are 14 hours earlier than New York City. Accommodations are private motel style suites. They hold a maximum of 14 people at a time, but that is only when full. Privacy and personal care here is at the highest level.

All Life Sanctuary

800-927-2527 ext 00205. Fax 501-760-1492. PO Box 2853,
Hot Springs, AR 71914. http://www.naturalUSA.com/retreat/

Viktoras Kulvinskas was co-founder of the Hippocrates Health Institute and is a pioneer of living foods preparation and wheatgrass juice therapy (*see Pioneers*). In 1998, he opened a camp and school for spiritual refreshment, lifestyle education, living foods preparation and wheatgrass on 100 isolated acres in the magical heartland of Hot Springs, Arkansas. The program here consists of classes in live foods culinary training, micro-agriculture (indoor gardening of sprouts and grasses), fermented foods, detoxification, digestion, food combining, lifestyle consultations, and self-evaluation using iris, palms, face, and tongue. There is a big emphasis on food here. Viktoras is a creative genius who can take grains and seeds and turn them into patés, dressings, burgers, cookies, crackers, chips, candy, fruit leather, granola, etc. You'll learn from the master.

Here, you have options to structure a program that is as intensive or relaxed as you desire. There is, for example, plenty of swimming, yoga, pranajama (breath yoga), aerobics, weights, trampolining, and options for hikes, canoeing, massages, and rest. Diagnosis is through physiognamy, kinesiology, astrology, numerology, iridology and reflexology—powerful tools that when cross referenced present a holistic map of your health and being.[3]

Grow Your Own

Photo by Sonoma Index-Tribune

Grower Richard Kersh at the Sonoma Valley Calif. farmers market

This chapter is about wheatgrass, not wheat grass. As explained in the primer on grass (*see p. 4*) wheat grass is grown outdoors in the field and wheatgrass is grown indoors in trays. There are many differences between these two which have to do with the growing time before harvest, the jointing stage (*see p. 53*), light and other growing conditions and the nutritional complexion (*see p. 52*). But since most readers are not farmers, indoor gardening is the most practical way for you to grow your own grass. This chapter tells you how to do it. It includes both soil and non-soil methods, growing tips, advice from experts, and how to make your own hydroponic grower.

Soil Method—Collect the Necessary Materials

Seeds & Trays. Purchase organic hard wheat berries, either winter or spring, from a natural food store, from a mail order sprouting seed supplier or your local wheatgrass grower. (*See Resources.*) Purchase growing trays from any garden store. They are called seeding trays. A standard size plastic seedling tray is about 11x21 inches. Choose the kind without holes in the bottom, since you don't want any leaks in your house. If your retailer doesn't have it, they can order it. Different size seedling trays exist. If you prefer larger or smaller trays, simply modify these instructions accordingly. Set up some shelves in your growing room according to the guidelines on page 136.

Use light and airy soil.

Soil. The best soil is light and airy with approximately 50% peat moss, 40% organic top soil and 10% vermiculite, pearlite or other fluffy soil aerators. This is a typical deluxe potting soil mix available at garden stores, but anything close will do.

Sprout bags are the easiest way to get your seeds started.

Sprouters. A sprouter is recommended to get your seeds started before planting. Sprouting bags are preferred over jars and are available at a few natural food stores or the mail order suppliers listed in the *Resources* chapter. Bags breathe and drain better than jars in addition to taking up less space. There is also less handling on your part, since you need only dip once daily to moisten them. You'll save space too, since they can hang instead of filling up your shelf, dish rack or counter. Nevertheless, jars have been synonymous with sprouting for decades and you can certainly use them. But if you're gearing up for some serious growing, you will find your sprouting chores easier and more efficient if you obtain a few sprout bags.

Basic Steps for Growing Grass in Soil

1. Soak 2 cups of grain for 9–12 hours.
2. Sprout for 2 days, rinsing twice per day.
3. Lay seeds on top of soil.
4. Water with a sprinkler or shower.
5. Cover the seedlings
6. Set seedling tray in a shady spot.
7. Check daily and moisten if necessary.
8. Remove cover when 2-3 inches tall.
9. Expose to light; water daily.
10. Harvest in 9-12 days or 6-10 inches tall.

1. Place 2 cups of grain in a jar and soak in pure water for 9–12 hours.

2. Drain out the soak water and rinse the seeds well in fresh water. Germinate the seeds for two days in a sprouter rinsing at least twice per day.

3. Fill your seedling tray with 1–2 inches of soil.

Lay the sprouted grain evenly on top of the soil one level deep.

4. After 2 days of germinating in the sprouter, the seedlings are ready for planting. Healthy seedlings will have a single thick shaft emerging from the grain and several white hairlike rootlets. Lay the seedlings on top of the soil spreading them evenly. Two cups of dry grain fills a standard 11x21 inch seedling tray. But if you are using a different size tray, simply lay out enough seeds to cover the surface one level deep.

5. Water the entire tray with pure water using a watering can that has a sprinkler head. The sprinkler is important in order to get a gentle and even shower. Water only enough to moisten the soil. Be careful not to over water. There are no drainage holes in the tray. Excess water can fester and create mold.

Water only to moisten the soil then cover with a second tray.

6. Cover the entire tray by laying a second tray on top. The top tray can be inserted in a nesting fashion or inverted and placed on top sandwich style. Alternately, you may cover it with a dark plastic garbage bag. Either keeps moisture in and light out. If using a bag, tuck it in loosely around the edges. Don't make it air tight.

7. Set the tray in a shady spot away from extremes of cold, heat, or wind. Wheatgrass grows best in cool weather.

8. Check your crop daily to make sure the seedlings are not drying out. Water only if they're dry. A mister or atomizer is ideal at this point for adding a small amount of water to the seedlings without significantly wetting the soil.

When 2-3" tall, expose to light and water daily.

9. When the sprouts reach approximately 2 inches in height (about 3 days), start watering with your watering can. Water once daily, being careful to use only enough water to moisten the soil. If you were to hold the soil in your hands, it would feel wet but not drip.

At the 2" height, the plastic lies loosely over the sprouts. Or, if using a second tray to cover, it would be inverted sandwich style. At 3", eliminate use of any cover and expose the growing grass to normal light.

10. Wheatgrass is mature and ready for harvest when 6–10 inches tall. This usually falls between 9 and 12 days depending on the climate. Another marker is when the blade develops a second stem. Harvest by cutting about one inch above the soil using a serrated knife.

Harvesting

Growing Tips

Soil

Quality of soil is critical. Most pre-mixed potting soils available at garden center stores are satisfactory. Deluxe potting soils usually include larger amounts of peat moss and soil aerators like vermiculite and perlite. Light and airy soil works best. You can make your own soil from organic compost. Dr. Ann Wigmore is famous for having designed a method of composting indoors in a bucket with worms yet having no smells! Compost increases the nutritive value of the soil and thus your grass. You can start two growing trays side by side from the exact same seed, same growing conditions and the grass with composted soil will be taller, greener, fatter, and juicier. The secret is in the compost. If you are interested in composting, obtain a good book on the subject. Although the subtleties of composting can get quite technical, the process is fairly simple. After all, the worms do most of the work.

Seeds, Seed Quality and Storage

Seed quality should be one of your highest priorities. The seed you choose may very well make the difference between an easy-to-grow, trouble-free crop and a harvest of mold and headaches. 'Organic' is not your only criteria. Good grass-growing grain is generally low in moisture

Troubleshooting Check List

1. Soil: lightweight and airy
2. Seed: test and change
3. Watering: moist but not wet
4. Air: breeze or slow fan.
5. Temperature: 70°F ideal
6. Humidity: keep low
7. Light: sunlight penetrates to soil
8. Even, indirect light
9. Location and environment
10. Season

and high in protein. For wheat, that is about 10% or under for moisture and about 12% or better for protein. Nevertheless, these numbers do not tell the whole story. The only way to know is to test. If you have some friends who grow grass, ask them for a recommendation. Order a small amount from different suppliers and make your own tests. Or, call your local professional grower and get his recommendation or buy seed from him. Good seed is golden.

Photo by R. Ross, Optimum Health Inst.

Two day old sprouted wheat laid evenly, one layer deep, on top of soil.

Grass can be grown successfully from hard winter or hard spring wheat. Soft wheat is not recommended. Kamut, the ancient Egyptian, non-hybrid durham type wheat, makes an excellent grass. The grain is generally 17–18% protein and makes mild tasting, thick grass blades. Green Kamut Corporation is a company that grows grass from this type of wheat in the high plains of Utah (*see The Companies*). Barley is a wonderful seed for growing grass and many claim that it is nutritionally superior to wheat. Green Magma is a very popular barley grass juice powder with much scientific research behind it. The seed however, is not available at any natural food stores and cannot be special ordered from them. Your only luck will be with the specialty mail order seed suppliers (*see Resources: Seed Sources*) who carry un-husked barley for the specific purpose of sprouting. Don't even think about testing the barley sold in health stores. Once you've got the right seed, you'll enjoy this slightly taller and broader blade. Oats are even harder to obtain than barley. Although animal feed stores will provide both, the high amount of debris in their product makes them unsuitable for sprouting. Rye and spelt are available in many natural food stores and similar to wheatgrass.

You may need to buy seed in volume, especially if you discover a good batch or, as with barley, it is hard to get. Store your grain in a cool dry place. Basements are usually good for coolness but bad for moisture. A perfectly sealed bucket is a successful way to prevent moisture and vermin damage. Don't buy bulk seed in the spring or summer. Grain is difficult to store in hot weather. Buy in the fall and in northern climates you will have nine months of winter temperatures for cool storage.

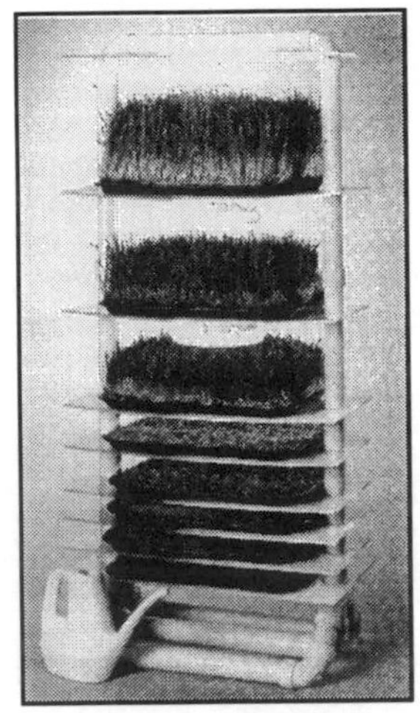
An 8 tray stand-alone grower from Sundance Industries.

Setting Up Your Growing Room

Your growing room must have convenient access to water, light and space. If necessary, you can choose to use indoor grow lights and a watering can. Shelves can be located on the window, on the wall or as a stand alone unit. Shelf brackets and shelves are readily available at hardware stores and are easy to install. Space the two vertical brackets so that each shelf is divided into three sections for three trays. Since each pair of brackets can support 3–4 horizontal shelves, this enables a 9–12 tray capacity. Stand alone units can also be constructed with four legs and can be 6 feet high fitting 16 trays. Visit your garden center store or get out your tools.

Light

Little or no light is required during the first four days of germination. Dark or shady areas are best. That is why the seedlings are covered with a second tray or dark plastic. After this time, place your sprouter in indirect sunlight, i.e. room light or a bright area. Avoid direct sunlight immediately after uncovering. Sun is both good and bad, just as with people. Sunlight on the soil keeps mold in check, but without proper ventilation, can raise temperatures to 100 degrees and cook your sprouts. Just one exposure of this heat is enough to start the process of rot, bacterial growth and mold.

If your location exposes your grass to several hours of direct sun, you will need to increase its diet of water. The more light, the faster and taller they will grow, but also the faster they will dry out. So monitor them carefully. Indirect sun provides more balanced temperatures and even growing conditions. Since there are fewer highs and lows in temperature and moisture, there are fewer problems.

Open a window to circulate the air or use a slow fan. If there are long periods of direct sun without a breeze, the grass may overheat. Relocate your crop if necessary. Three hours of sunlight per day is plenty. If you do not have sunlight, they can grow their entire life in indirect light, shade or with full spectrum grow lights.

Harvesting

Photo by Robert Ross, OHI

Just harvested!

Harvest time is going to be slightly different for every location. The general recommendation is 9–12 days. The grass will not be very nutritious if it is too young or if it starts to fall over and turn yellow. Although outdoor grass can grow much longer, the 2 inches of soil limit the root space and thus its longevity. Better to cut the grass and refrigerate before it falls over than to leave it in the tray. You can also judge maturity by the height of the blade and by observing when the blade splits and develops a second stem. Don't, however, confuse this with the jointing stage. Tray grown grass does not have enough soil to joint. If you grow your grass in a terra cotta planter with 10 inches of soil, you can achieve some of the benefits of outdoor field-grown grass. (*See Jointing p. 43.*)

Pure Water

Water is your primary ingredient. It is the river on which all life flows. If this grass is your medicine, you surely want to use the highest quality water. Purchasing bottled water is uneconomical because of the volumes involved. You should be drinking, cooking, sprouting and washing vegetables in pure water, so a home water purifier is the best choice. Distilled water, loose granular carbon filters, hard carbon block filters and reverse osmosis are your options. Since plants thrive on mineral content, you can add liquid kelp or another source of minerals to any of these waters to enrich them. This is especially valuable for hydroponic gardening and anytime you use distilled water since it is mineral-free.

Mean Mr. Mold

Molds, fungi, and other micro-organisms are a part of gardening and food, especially when no chemical fungicides, insecticides, or mold inhibitors are used. Mold on wheatgrass is generally a function of temperature and humidity. Some professional growers maintain that 70°F is the ideal growing temperature. Humidity is another scourge and a good reason for air conditioned growing spaces in places like Florida. Mold often starts out as white cotton and turns gray as it grows (just like people). Many mold emergency calls are false alarms. Young grass has cilia hairs which are microscopic fringe on the rootlets. As drops of moisture cling to them, they look very cottony. The cilia will disappear with watering. Mold doesn't or will bounce back quickly.

According to master grower Michael Bergonzi at the Optimum Health Institute, seed is crucial to a crops' propensity to mold:

Seed is like people. Some people get sick easily, others don't. Two people are in the same room exposed to a virus; one gets it and the other doesn't. The better seed is more immune or resistant to mold. Mold is always in the air; it just depends whether or not the seed is strong enough to fight it off.

Sometimes mold is a process of elimination. You change the trays, change the water, soil and, if nothing works, it comes down to the seed. Hydrogen peroxide is a natural, non-toxic mold killer. Spraying or misting the soil with it is helpful, but it is still a bandage measure.

Nagging Gnats. Flies show up when there is mold because mold is dinner for them. They won't hurt the grass, but they are annoying. Good air circulation with an exhaust fan is beneficial. Vacuuming the buggers off the trays also works. If you increase the air circulation and lower the temperature, most flies and gnats will disappear. If you have an infestation and it is winter, clean out everything and chill the growing room for a few hours. That will give you a fresh start.

Storing Fresh Cut Grass

Sprout bags, glass or plastic containers with lids and green eco-storage bags are the best way to extend the life of your refrigerated grass. Standard plastic bags are not recommended. They suffocate grasses and vegetables. Sprout bags are natural fiber sacks that breathe and drain on all sides. Unlike plastic bags, they don't create puddles of water that breed bacteria and mildew. Flax or hemp fiber is the best material for these bags since they remain moist. A plastic or glass container also allows air circulation because it does not shrink or conform to the sides of the grass or vegetable

Photo by Robert Ross, Optimum Health Inst.

Mature wheatgrass ready to harvest.

and has airspace at the top. Green eco-bags are chemically different from standard plastic bags and measurably increase the longevity of your refrigerated crop. Grass cut in its prime and refrigerated in this manner lasts for approximately two weeks.

Don't eat yellowish grass. Juice your grass before it starts to turn. If you have more grass than you can use, juice it and pour it into ice cube trays. One day when you are out of grass, you can drop a grass ice cube into a carrot juice and enjoy it.

Interview with Master Grower Richard Rommer

Richard Rommer grew wheatgrass for Dr. Ann Wigmore as early as 1973 while apprenticing at her Hippocrates Health Institute in Boston. His company, Gourmet Greens, is one of the few that ships fresh grass and soil-grown sprouts via refrigerated next day air in the USA. (See Resources.) *Questions in brackets [ ] are the author's.*

Master grower Richard Rommer

[*What's the most important thing you have to tell home growers?*] Sooner or later all wheatgrass growers encounter mold. It is a white or gray cotton looking growth at the soil level that if left unchecked will stunt the wheatgrass and even prevent the leaves from turning fully green. It is especially prevalent in the summer during hot humid days. There are several things the grower can do to keep mold to a minimum.

First, use good soil that is fully composted. Organic matter is great but if it is not fully broken down by beneficial micro-organisms, it will contribute to the buildup of mold. If you are growing with a store bought soil mix, make sure it is free of anything like bark or bits of hay or straw. It should be consistent and free flowing.

Second, get good seed. Many people start growing with a small bag of wheat purchased at a health food store and wonder why they can't grow good wheatgrass. That seed may be old or improperly stored and no one has tested it to see if it grows good wheatgrass. There is an abundant supply of good quality organically grown seed out there. Buy your seed from a wheatgrass grower or sprouting seed house that has tested it for their own use.

[*Any tricks on getting the seed started?*] Soak your wheatgrass seed in the refrigerator for 24 hours whenever you are having mold problems. This prevents the soak water from getting funky. And when you pour off the soak water, rinse it with plenty of clean water until all the colored soak water is gone. Let the seed sprout for about 12 hours before you spread it onto the trays.

We stack our trays with the bottom of one tray directly on top of the one beneath it for 48 hours at 67°F. At the end of the 48 hours, the trays are separated and transferred to the growing shelves.

[*What about water and light?*] Water immediately but do not over-water. Water only once a day. After one day, your trays should not be dried out—but they should need water. One half hour after watering, the soil should have the moisture of a wrung out sponge. That is, if you take the soil into your hand and squeeze gently no water should drip from your hand. But if you squeeze tightly, water will drip out.

If you live in a sunny climate, a greenhouse is best. Second best is good grow lights. The young shoots of the wheatgrass will follow the light. Good light getting down to the soil level will help to keep mold to a minimum. Harvest the grass before it begins to lose its vigor and fall over on its own. Once it falls over, you are no longer getting light down to the soil level.

A slow moving oscillating fan will keep the air moving. Direct the air flow directly onto the trays. This may tend to dry out the soil in the trays so keep an eye on the soil moisture. You want the young sprouts to have plenty of available water but they cannot be sitting in pools of water. Remember the wrung out sponge test.

Harvest the grass just before it starts to fall over. If it is more than you can use, store the extra wheatgrass in a plastic bag in the refrigerator at 34–37°F. Fresh harvested grass is best but don't think that yellowing, falling over grass is better than refrigerated wheatgrass. [*End*]

Soil-Free Hydroponic Grass Growing

Traditionally, all wheatgrass is grown in soil. All the major wheatgrass healing resorts grow it that way and Dr. Ann Wigmore, the mother of wheatgrass, never supported any alternatives. However, none of the resorts oppose the growing of wheatgrass hydroponically. Robert Nees, director of *Optimum Health Institute*, reports that once upon a time, OHI grew and served hydroponic grass to its guests. He says "There is only a minor difference in nutrient value between the two grasses. We only stopped [growing hydroponic] because it was inefficient for us on a large scale operation." Some of the resorts sell home hydroponic growers in their on campus stores. Dr. Chiu-Nan Lai, of the University of Texas System Cancer Center, used soil free wheat grass in her grass research and still reported that "the inhibition of activation of potent carcinogens is quite strong at a reasonably low level of extract." (*See Research*)

Soil vs. Non Soil

Soil is the best medium in which all green plants grow. But sprouts are traditionally grown without soil and wheatgrass is, after all, a mature wheat sprout. Some experts claim that the immature roots systems of the young sprouted wheat are not capable of absorbing most of the minerals in soil anyway. Hydroponic grass is milder in taste. If you like, you can also juice the whole grass, roots and all, which adds additional nutrients. Let's face it, growing grass in your home is easier if you don't have to deal with soil. There is simply less to do. But hydroponic grass is generally not as tall as soil-grown grass. Thus, soil-grown grass is still superior and the most dependable when growing for a serious health challenge. Choose the method that best suits your circumstances. Either way, you're a winner.

The method that follows differs from conventional hydroponics because it does not use non-soil mediums like sand or synthetic liquid fertilizers. For the most part, this method is just seed and water.

Hydroponic Grass Growers

Like all sprouts, wheatgrass will grow without soil. But, because of its copious and fibrous root system, ordinary sprouters won't do. Consequently, only a handful of soil-free wheatgrass dedicated sprouters are on the

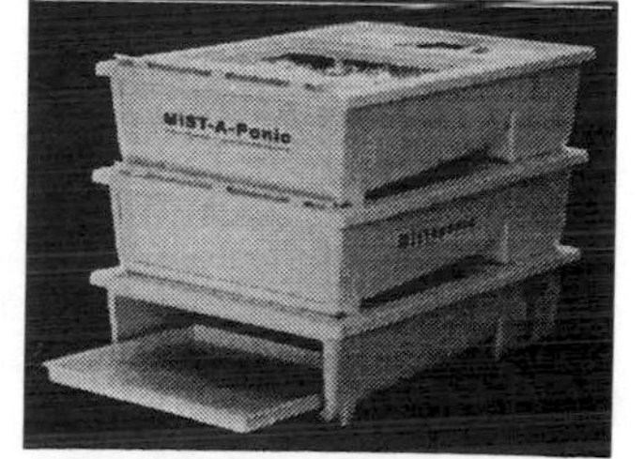

3 Tier Mist-a-Ponic grass grower

market. Fully automated, stainless steel professional units cost well over $1,000. One affordable automatic unit is made of plastic and has the capacity for up to three levels so you can cycle your crops providing a continuous grass supply.[1]

You can also, without much difficulty, make your own. The following are instructions on how to make a hydroponic grass grower and the steps to grow grass successfully in it. Many of the considerations about light, water purity, temperature, etc., discussed earlier in this chapter, also apply for hydroponic grass. If you do not want to make your own grower, contact the vendors in the *Seed Sources* and *Mail Order Shippers* section of the *Resources* chapter to learn what sprouters they have available. The author has designed an under $30 sprouter similar to the 'home-made' model described here. It does a good job on wheatgrass and sprouts and is commercially available from some natural food stores and mail order suppliers.[2]

Make Your Own Hydroponic Grass Sprouter

First obtain plastic seedling trays from your garden center store. Choose the kind with holes in the bottom. The seed amount in these instructions applies to a tray approximately 11x11 inches. Adjust the seed amount if your tray is different. Secondly, find a heavy weight (4 mil thick) plastic bag or flat plastic sheet available in the paint department of your hardware store. Fold, cut and tape the plastic into a tent that can sit over your seedling tray. Airspace above the tray should be approximately 10–14 inches. Leave enough length to partially tuck it under the growing tray.

Homemade greenhouse tent

1. **Soak Seeds.** Choose hard winter or hard spring wheat berries, sprouting barley or Kamut. Soak 5 tablespoons of your chosen seed in 8+ ounces of pure water for 9–12 hours.

2. **Germinate.** Sprout the grain for 2 days or until the shoots are about twice the length of the berry and there are several long, hairlike roots.

3. **Set.** Pour the seeds into the tray and hold at an angle to allow excess water to drip out. A few ungerminated seeds may pass through the holes at the bottom. Some seed loss is normal and insignificant.

Insert the sprouter tray into the soft plastic greenhouse tent, tucking it loosely underneath. Do not attempt to make it airtight. Place the sprouts in a shady or mostly dark spot where the temperature is, ideally, 65–75°F.

4. **Grow and Rinse.** Rinse twice per day (or more) by removing the tray from the greenhouse tent and running it under a strong shower of cool water. The sprouts prefer to be showered rather than drilled with the jet of a faucet. You may already have a dish spray hose that is built into your sink. Use it or buy a faucet spray adaptor from the hardware store. Misting with a spray bottle is verboten. Misting does not provide the necessary water pressure to wash away waste and bacteria.

The purpose of rinsing is to: a) cool down the sprouts and, b) wash off waste materials such as mold and fungus. Adequate rinsing eliminates mold and most common sprouting problems. Spend 30 seconds hosing your sprouts with a strong shower of water.

Immersion. After about the fifth day of germination, an alternate method of rinsing called immersion can be used. Remove the tray from the greenhouse and immerse it in a sink full of water. If your sprouts have anchored themselves into the holes at the bottom, they will not fall out of the sprouter. At the same time, the sprouts are getting a total bath and thus a thorough rinsing. If you have many holes in your tray and your crop is securely rooted, you can turn it upside down and gently swish it back and forth. This is a great technique for getting rid of hulls from sprouts and is a simple, fast and thorough method of rinsing.

5. **Harvest.** Harvest time for wheatgrass is when the blades have reached their richest green, usually between 9 and 12 days. They should be approximately 6 inches tall when grown hydroponically. To harvest, grab a small group of blades and carefully jiggle them out of the sprouter.

About the Greenhouse Tent

The purpose of a greenhouse is to: a) retain moisture; b) maintain warmth; c) allow the penetration of light. This homemade greenhouse, made of heavy-duty, soft plastic, is perfectly clear, sturdy and allows the entrance of beneficial ultraviolet rays. House the seedling tray in the greenhouse at all times for optimum results. Place the sprouter inside and tuck the open end loosely underneath. Remember, it does not have to be airtight. Create an envelope of air by mounting the greenhouse so it looks like a tent. It should be strong enough to stand on its own. At full height, your tent should be about 12 inches tall. This provides valuable airspace for the sprouts to breathe. During the first 4 days of germination, the small seedlings do not require the full extension of the tent since their respiration is so shallow. You may puff the tent up to half height during this time. Growing without the greenhouse is risky and not recommended because the sprouts can dry out, creating stunted growth and mold.

Growing Other Sprouts with This Method

Green pea shoots, buckwheat and sunflower greens also grow successfully following the previous instructions. Green pea shoots grow 10 inches tall in approximately 9 days. You can use either green or red peas. Buckwheat takes approximately 12 days. Use only the black, unhulled grain. They are mature when at least 80% of the black hulls have fallen off and their clover-like leaves have unfolded. Sunflower also takes about 12 days and is ready when 80% of the shells have dropped, revealing a hearty V-shaped sprout. To harvest, grab a small bunch and gently wiggle them free or cut them just above the roots.

Cost Comparison: Buying vs. Growing

	1998 Purchase Price per bottle or tray	#Servings Serving Size	Cost per Serving
Purchase Grown Grass[1]	$12.50 lb	10	$1.25
Barley Grass Juice[2] Powder	$35 @5.3oz	50 @3g	$0.70
Wheat Grass Juice[3] Powder	$57 @8oz	76 @3g	$0.75
Whole Grass Powdered[4]	$56 @24oz	227 @3g	$0.25
Juice Bar Grass Juice	$1.75 @1oz	1 @ 1oz	$1.75
Home Grown & Juiced	$0.80/lb/tray	10	$0.08

1998 prices. 1. Mail ordered shipped grass. 2. Green Magma. 3. Pines Juice powder. 4. Pines whole leaf wheatgrass powder. Prices will vary, but the relative costs should hold true.

How Much Does It Cost?

There is no question that growing your own grass is more economical than having someone else grow it for you. But the low cost of getting juice from your own crop may be somewhat deceiving. The hidden cost is labor and equipment. The price of your time, materials and juicer make the juice bar a viable alternative and one that is not as expensive as it first appears. Unfortunately, fresh wheat grass juice from juice bars or health stores is not available everywhere. If you are unlucky with your location, you could dial a 1-800 number and mail order your grass for tomorrow morning. The Resource chapter lists several vendors that offer this service. That eliminates the growing, now all you have to do is juice. But this may not be cost effective if you are a volume user. You could hire a gardener or, lastly, you could purchase store bought grass powders, either whole leaf or juiced. The convenience and portability of the powdered juices is a most satisfying way for Americans to get healthy. However, the cost of growing, juicing, drying, and packaging is dear. These production costs are simply not present in growing your own fresh grass. When all the calculations are done, it turns out that soil is still dirt cheap—and so are seeds. At only 8¢ per ounce, you can guzzle enough of your own home-grown grass to keep in-the-green for a long time.

Photo by Robert Ross, Optimum Health Inst.

Eight inch tall mature wheatgrass.

The Juicers

This chapter introduces you to some of the equipment necessary to make your own fresh squeezed grass juice. Since your health depends on it, you will want to choose carefully and explore all the options. Across the industry, manufacturers of grass juicers generally produce good machines. You can't go wrong with any of the juicers described here. But still, there are many styles and features to be considered in order to best match your circumstances. It is not the job nor the goal of this book to compare, only to explain and introduce.

This is not a list of every grass juicer available, but it definitely has the major players. Pricing is quoted in terms of 'ranges.' The range identified for these juicers is a general point taken from the manufacturer's suggested retail price in 1998. No one needs to tell you that prices change over time. The marketplace also offers different opportunities such as wholesalers and discounters, so you may very well pay less than the suggested retail price. The author has no relationship, financial or otherwise, with these machines or their manufacturers. No preferences are stated or implied even by the space allotted in discussing them. This chapter merely serves to introduce you to an essential part of the program: the grass juicing machines. Note that grasses are also available in powder and tablets which are an alternative to home juicing. Read the *Nutrition* and *Healing with Grass* chapters for more information on the differences between fresh and dried and how to decide which grass is best for you.

Miracle Manual wheatgrass juicer

Since wheatgrass juicers are basically grinders and presses, many of them also make nut butters and purées. This includes such delights as peanut butter, almond butter, sunflower meal, sesame butter and frozen desserts. Attachments for making flour and pasta are also available.

Why Common Juicers and Blenders Won't Work

The wheatgrass juicer is actually a screw press. Unlike conventional vegetable juicers that tear or cut vegetable fibers into tiny fractions, the grass juicer presses and squeezes the pulp, much like wringing the wetness out of your laundry. Conventional juicers are simply not designed to handle the woody, ligneous fibers of grass and, besides, they operate at such high speeds, sensitive enzymes would oxidize immediately. Regular juicers destroy the nutritional value of grass, but that's not all. They frequently smoke, choke and otherwise dance around the counter damaging themselves. And, when it comes to juicing leafy vegetables like parsley, spinach, lettuce, celery, herbs and sprouts, common high speed vegetable juicers are no match for the slow moving grass juicers

Blenders are also unworkable. The green blades will wrap around the steel blades strangling your blender to death. Even if it survives, the high speed blades aerate and heat up the grass, and don't make much juice anyway. Even the famous Vita-Mix, the all powerful multi-purpose kitchen appliance, is neither designed nor recommended for juicing grass.

The Green Power Juicer

888-254-7336. http://www.greenpower.com Downey, CA 90241.

When the *Green Power Juice Extractor* came along in 1994, it instantly became the most exciting entry on the juicer playing field in decades. With its patented invention of a twin gear system, it is both innovative and unique. It is also avant-garde. Inside the stainless steel gears are magnets and ceramics that produce positive ions as the gears spin. According to the designers, this effect results in less oxidation and greater longevity of the fresh juice. The company provides the results of an independent lab study evaluating carrot and apple juice over 72 hours. Theoretically, the juice lasts longer because more enzymes and nutrients are kept intact, enhancing stability. Users claim that carrot juice can be refrigerated for 48 hours. This saves the chore of juicing each time you want to drink.

One of the main advantages of this juicer is its versatility. It will juice carrots and wheatgrass at the same time. Traditionally, these two required different juicing machines. Juicing carrots right along with your wheatgrass is wonderful for both taste and convenience and may convert even the most reluctant wheatgrass drinkers. It also juices sprouts, herbs, grasses, leafy greens, apples and a wide variety of vegetables and fruits. Use this same machine to entertain non-wheatgrass guests with nut butters, frozen fruit sorbets, mochi (Japanese rice cakes) and 3 different shapes of pasta.

The twin gears turn at a slow 90 rpm's, crushing the vegetables and forcing the juice out against a fine sieve. Since it is also a pulp ejector, you can juice continuously without stopping to clean. When you do need to clean up, there are four parts to wash and remount: the twin gears, strainer and front and back housing. This is no more or less than setting up and cleaning a common vegetable juicer. This is also one of the quietest machines, eliminating the maddening noise, heat and vibration that has traditionally been a by-product of the juicing process. Whoever designed *Green Power* was a creative individual and a dedicated juicer. This shows in the magnetics, the one-machine-juices-all design, the multi-purpose appliance functions, the low noise, and the sturdy built-in handle for moving and storing. No surprise it won the 1993 Silver medal prize in Germany at the International Exhibition of Inventions. Price range: $600.

Omega Twin Gear

800-633-3401, Fax 717-561-1298. PO Box 4523, Harrisburg, PA 17111

This is the newest entry into the juicing market, making its appearance in 1998. It is a twin spiral gear juicer that does both common fruits and vegetables like carrots and apples in addition to grasses, sprouts and leafy green vegetables. *Omega* is a consumer oriented company and their original vegetable juicer is highly rated for customer satisfaction. This is also evident in the new *Twin Gear* model which comes with numerous 'creature fea-

tures' to make your juicing easier. You get three cleaning brushes and sticks, a vegetable pusher, a generously sized funnel, a cup to capture the fresh juice and even a pulp residue container. You won't have to search around the house for cups and saucers that inevitably don't fit. All this makes for neatness, compactness and convenience. You've got five years of peace of mind, too—one of the industry's best warranties.

Assembly involves only three primary parts: the stainless steel twin gears, the strainer and the cover housing. All are neatly held together with an easy-to-use clamp. Because it's a pulp ejector, you can juice all day without stopping to clean. Since all vegetables are different sizes and textures, traditional juicers tend to do better on some vegetables than others, but this machine has a neat feature. A knob at the front widens or narrows the gap between the gears. This means you can adjust it according to what you juice. That really helps for sinewy fibers like wheatgrass and ligneous greens like kale. There is even a great escape feature in case you overstuff your machine—a reverse button. Because wheatgrass juicers are slow turning machines (centrifugal juicers turn at 3,600 rpm's), they can get bogged down with compacted pulp, occasionally forcing you to disassemble the machine. But the reverse and adjustment features will get you out of a jam.

At 2 horsepower, this is one of the most powerful motors around. That's plenty of power for volume users and the whole family. Its slow 90 revolutions per minute won't create any heat as it chews the fibers, penetrating plant membranes and releasing live nutrients. The pulp on the juiced grass is very dry and even that is adjustable via the front knob. Of course, since it is also a pulp ejector, you can simply reinsert the pulp to squeeze out even more. Price range: $500.

Opti-Fresh Wheat Grass Juicers

619-464-3346, Fax 589-4098, 6970 Central Av, Lemon Grove, CA 91945

Opti-Fresh All Stainless steel juicer.

These are the only all stainless steel juicers made today. They are worm gear, screw-press machines that stem from the original grass juicer design by Ann Wigmore. These are dedicated wheatgrass machines. No carrots or apples are possible here, however, you can easily juice all leafy greens and sprouts. While the original machines looked similar to the old cast iron - grinders, *Opti-Fresh* has moved beyond tin and iron into stainless steel and enamel. The distinct advantage comes in clean up. Iron machines rust and need to be oiled and 'cured.' This is just one more chore for the user and a tricky one at that. In the worst cases, iron and oil leak into your juice. Although in Dr. Ann's day users negotiated around these problems, *Opti-Fresh*'s machines are rust-free and easy-to-clean. They are also heavy duty affairs with 1/3 horsepower motors capable of juicing all day. They're powerful, yet stable and quiet. These juicing machines are manufactured for the *Optimum Health Institute* (see *Healing Resorts*). They are the machines that guests use for their healing. Four ounces of juice takes just 2–3 minutes. Price range: All Stainless Steel $750. Enamel and Stainless Steel $700.

Sundance Wheateena Wheatgrass Juicers

914-565-6065, fax 914-562-5699. PO Box 1446, Newburgh, NY 12551

This is the original wheatgrass juicer company that pioneered Dr. Ann Wigmore's idea into a standard design that became the prototype for all wheatgrass juicers for 30 years. Back in the middle 1960's it was a patching together of a motor and a modified berry press made by one of the few remaining American cast iron foundries. These worm gear, screw-press workhorses crushed and squeezed berries and nuts for generations. The addition of a stainless steel screen enabled them to capture the juice from grass. The pressure can be adjusted via a screw on the front to make your grass drier. The company also makes a hand cranking portable juicer that clamps onto any steady table, and four other electric models. All units are cast iron with heavy duty, hot dip tin plating (except in grinding areas) to prevent rust and make a durable finish.

The Wheateena Workhorse Juicer

Their most popular consumer model is the *Marvel* which comes in two versions: high speed 157 rpm's and low speed 105 rpm's. These are solid machines mounted on a sturdy stainless steel platform and guaranteed for 3 years. They produce about an ounce per minute. Price range: $300/350. Their *Wheateena Workhorse II* juices about 2 ounces per minute. A grinding mill attachment is also available to make flour. It turns at 91 rpms but has a whopping 130 inch pounds of torque (a lot of twisting power). Price range: $600. The *Sundance Manual Wheatgrass Juicer* is a ten pound precision made cast iron piece that includes a strong clamp. You can do about one ounce per minute with a manual but, because of the effort required, they are only recommended for 2–3 ounces per day. Price range: $200.

Miracle Wheatgrass Juicers

Miracle Exclusives, Inc., Port Washington, NY 11050. 800-645-6360, Fax 516-621-1997. Miracle-exc@juno.com

Miracle Exclusives is an importer of three wheatgrass juicers. All are clean, modern kitchen appliances imported from Europe. They have the reputation for being the most affordable juicers in the industry. Their motor housings are white plastic as is the juicer section. The plastic auger screws against a stainless steel sieve squeezing all the wheatgrass out. These machines are auger, screw-press designs similar to the cast iron machines, except they trade iron in for plastic. Their MJ550 model is the most widely sold electric wheatgrass juicer on the market. Price range: under $200.

Low cost Miracle Electric model 550

Their professional model is made of finely finished cast iron with two stainless steel strainer screens. The motor housing is a hard enamel finish with a carrying handle for easy transport around the kitchen. Optional attachments are available for grinding flour, beans and nuts. It has two motor speeds, a huge hopper, and is very, very quiet. Price range: $500.

Their manual juicer is the best selling manual wheatgrass juicer in the world. It is a heavy duty cast iron machine with a shiny hot dipped tin plating finish. It is sturdy and compact and even includes a wooden handle on its crank. Price range: $75.

Photo Courtesy Green Foods Corp.

Greet the Products You'll Meet on the Store Shelves

Choose the grass products that best fit your diet and lifestyle

This chapter will introduce you to the companies you'll greet on your health food store shelves and in catalogues. It is not intended to be a complete list of all manufacturers. But it is indeed a thorough list of the producers who grow and process the grasses that end up in the products you see. For additional listings, see the *Resources* chapter and visit the website for this book at http://Sproutman.com. The goal of this section is to help you make informed purchasing decisions that are suited to your diet, health and lifestyle needs. No recommendations are intended[1] and the product listings for each company do not reflect their entire line. Neither the author nor publisher receive financial benefit from any of these companies. But the author has used all of the products and visited, met or spoken with each company. The background information below is intended to give you a sense of the company behind the product so you can match your inclinations to theirs. You cannot go wrong with any of these products. And if you are a connoisseur, you might try them all! You are sure to get healthier with each one. Remember, the goal is to find a convenient and efficient way to include grass in your daily diet.

The Field Growers

Field growers have hundreds or thousands of acres to manage. They cut their grass at a low 8–10 inches and at the jointing stage that Charles Schnabel defined. If temperatures are warm, this period can progress quickly and manufacturers are hard pressed to harvest their crop in a short time. However, they must meet the task or lose their investment. Growers must be devoted to capturing the plant at this peak time of maximum nutrition and then protect and maintain it right through the harvesting, drying, milling and packaging. If this wasn't hard enough, field growers are forever challenged by weather. If it is too hot, too cold, too rainy or too dry, there is trouble. It is not unusual for a farmer to lose 50 acres of beautiful wheat, just because the ground was too wet to send out the heavy harvesting machines. By the time things dry up, the grain may have matured beyond the jointing stage and it's too late. As much as two hundred days of growing can be lost! Thanks farmers, for your hard work and valuable contribution.

Green*Life*™

V.E. Irons. Sonne's Organic Foods, Inc. Cottonwood, CA and Kansas City, MO. 800-544-8147, fax 916-347-5921

"No one seems to realize that to produce Greenlife we use more fuel for heating, condensing and refrigeration than most large hospitals. All this is necessary for one purpose only —to retain the life in the product." —V.E. Irons

This is the company that health pioneer V.E. Irons founded in 1946 and is the oldest existing grass foods producer in the world. (See *The Pioneers*.) Their product, **Green*life***™ has been marketed since 1951. It is the dried extracted juice of organically grown cereal grasses: barley, oats, rye, and wheat. The grasses are grown on 530 acres along the Gasconade River on rich Missouri soil, in fact in "Richland," Missouri. The grasses are cut just as the first joint emerges to maximize it's nutritional content. (*See Jointing p. 43*) Processing takes place at the same location. The grass is ground, the juice is extracted and chilled. In this cold state, the grass juice is then dried by a roll-vacuum process at 90°F. The dried grass comes out like crepe paper and is then ground into a fine powder. It is necessary to keep the humidity at 30% or below, or else the powder will absorb the moisture and turn it brown and gummy. To prevent this, their powder room has 6 to 8 inches of insulation on the walls to keep the outside humidity from entering. From the powder room the product is kept refrigerated at 38°F until it is placed into dark brown bottles with rubber gaskets in the covers. The loosely covered bottles are placed into a nitrogen chamber where the oxygen is drawn out of the bottle and replaced with nitrogen.

Products

This is an old fashioned "health" company. Their products are not fancy, not commercial in design, nor conventionally marketed. They promote health first and the products are simply the tools for health. V.E. Irons is famous for his philosophy of intestinal cleansing to achieve optimum health. All of this company's products fit under the theme of detoxification and intestinal cleansing. Its flagship product is a 7 day cleansing kit with psyllium seed bulk fiber and liquid bentonite detoxificant. Green*life*™ (also part of this kit) fits perfectly into the colon health regimen because it bathes the intestines with chlorophyll and nutrition. Approximately ninety-five percent of it is grass from barley, oats, rye and wheat. The remainder is beets and sea kelp. This is a "juice powder" with the grass fiber removed. Like all manufacturers in the industry, it is available as both tablets and powder. They also manufacture a "fortified" version which combines Green*life*, 60%, with fish liver oils, brewers yeast, papain, bone meal, and lecithin (not a vegetarian product).

> *"It would be so much easier to dry the grass in its entirety, pulverize it and sell you hay in ground form, or perhaps tray dry and spray dry the grass juice. The analysis might be the same but there is no life!. Remember, LIVE foods produce Live Tissue —Dead Foods produce Dead Tissue."* —V.E. Irons

The Pines Truck: "Feed the Hungry, Your support makes all the difference in the world."

Pines Wheatgrass

Pines International. 1-800-MY-PINES, Fax: 913-841-1252. Lawrence, KS.
InfoService@Wheatgrass.com http://Wheatgrass.com

Cereal grass is easy to grow. Whether or not it provides high nutritional concentration, depends on where you grow it, when you harvest it and how you store and process it. —Ron Seibold

Pines is the largest wheat grass grower in the world. Although only started in 1976, Pines carries on the tradition of Dr. Charles Schnabel the original pioneer of grass, as a source of human food (see *The Pioneers*) and grows it in the same place and in the same way as he did. Its founders, Ron Seibold and Steve Malone are not really farmers or businessmen. The concepts that launched this company are mostly social and environmental and they still live up to those principles. Both were disappointed with the increase in chemical agriculture and the rape of the land by corporate developers. Growing cereal grass and selling it provided a way to earn money and preserve land at the same time. What's more, community could be developed around working the land which would provide pleasant residential communities. They saw Pines as an opportunity to stay connected to the land they grew up on, and work with some of their neighbors to develop a sustainable community. The name "Pines" was chosen because of the Pine tree image of "green" and "natural." Seibold and Malone started the company with $50 and $100 purchases of stock from their own pockets, employees and farmers. It is still cooperatively owned today by its original founders, stockholders, farmers, neighbors and employees—truly a grass roots company!

Today Pines owns 3,000 acres of certified organic rich Kansas land, one-third of which has been converted from pesticides and herbicides. The company has committed much of its after tax profits to land preservation projects. In addition, they have an International Hunger Relief program in which they donate the grass foods they harvest to the needy worldwide. Since 1991, over $3 million worth of cereal grass foods has been contributed to Native Americans and the homeless in the U.S., and to organizations that distribute food for famine relief in Africa, Central America, Asia, Korea and Europe. Because grass is such an efficient crop, they grow all they need on less than 1,000 acres. Other acreage is used to grow alfalfa, beets and rhubarb which they make into juice powders. The rest is hay, other organic crops and forest. Pines has also established a Wilderness Community Education Foundation on their land and constructed three energy-efficient demonstration homes, two of which are entirely underground.

How The Grass is Grown

Pines plants in the Fall and harvests in the Spring. They believe that the slow growth of the plant during the Kansas winter is the ideal climate for growing grasses. During the approximately 200 days in between, the grass develops deep mature roots which pull minerals from the rich soil and send it up to the leaves in the spring. Pines makes sure to provide those minerals by rotating the wheat fields with soybeans and red clover. Harvest time in April and May is short and feverish. According to the principles taught by Charles Schnabel, the Father of Wheatgrass, the grass must be harvested just before the jointing stage when the plant sends its resources up into the development of the seed head or grain. After jointing, wheat grows rapidly and its nutrient concentration drops precipitously *(see Jointing, p. 43)*. The grass is only about 7–10 inches tall but the root systems, buried well below the frost line, are enormous. This gives the grass the ability to survive through the snow and cold and bounce back to vivid life in the spring while most vegetable crops must be replanted.

Harvesting and Processing

Growing healthy grass on good organic land is not enough. The grass must be harvested, dried, ground, tableted, stored and packaged. Meticulous attention has to be given to each phase or the product will lose its bio-activity. Pines harvests using special custom-built harvesters with sterile collection hoppers. The machines cut only the top 3 inches or so from the 7–10 inch tall grass which is only harvested once. If they took a second cutting, grew it longer or cut it lower, they would reap greater volume and save production costs but at the sacrifice of nutrition. The remaining grass is thus left to continue to grow past the jointing stage to full maturity. Later the company will harvest and sell this organic wheat and barley grain. Meanwhile, the harvester never lets the grass touch the ground, nor is it touched by human hands. The grass is very quickly transported from the harvester to the nearby dehydrator building by truck in sterile bins. It is literally dried within minutes of harvest. Were it left lying on the ground waiting for delivery, it would bleach in the sun, oxidize and become contaminated. Their custom triple-bypass drum dehydrator dries the leaves in only 8 seconds with a core temperature that never exceeds 107°F. All this is done to maintain maximum enzyme activity and nutritional integrity.

> *"People are becoming more aware of the importance of dark--green, leafy vegetables, but the traditional American diet doesn't include enough of them. Our products offer a convenient way to correct that deficiency without having to change your whole diet."* —Ron Seibold

From the dehydrator, the grass is pelletized to cut down on the surface area exposed to the elements. The pellets are stored in huge, nitrogen-filled pharmaceutical bags. Each bag holds a ton or more grass pellets. The bags are stored in underground caves at 10°F until ready for tableting. Samples of every batch are taken to test for contamination by microbes and for nutritional value. When needed, the grass is removed from storage for processing into tablets. The milling or fine grinding of the grass is done at super-cold cryogenic temperatures so there is no heat to destroy product vitality. Tablets and powder are then packed in amber glass bottles in a vacuum chamber where all oxygen is removed and replaced by nitrogen. This enables long term shelf life and prevents deterioration and oxidation from exposure to air and light.

> I *pulled a test on myself to make sure the wheat grass was actually helping me. I quit taking it. In two months time, I can tell you that physically it made a difference. People have even noticed a difference in my skin and eyes. Some casual acquain-*

tances have even asked if I was ill. (No one but I knew about this self-imposed test.) [2]

Products and Fiber

Pines manufactures a full line of wheat and barley grasses in powder and in tablets. Their lead products are 100% pure whole leaf dried grasses of wheat and barley. No fillers, binders, or other ingredients are used. They also manufacture a "juice"powder in which grass is juiced and then spray dried onto arrowroot. Approximately 20% of this product is arrowroot, a tuberous vegetable starch that functions as a carrier for the juice crystals. Pines promotes the consumption of grasses as whole fiber vegetables instead of pushing juice powder. They proclaim the many proven benefits of vegetable fiber and the shortage of it in the American diet. Furthermore, fiber carries some nutrients which are lost when it's removed. Not only does their grass provide the benefits of fiber, but it is very economical. To give you an idea, their 24 ounce bottle of whole wheatgrass powder retails for $56 while their 8 ounce size of wheatgrass juice powder costs $57.[3] Nevertheless, they make both, including beet and rhubarb juice powders. Their combination product is called *Mighty Greens* and is a blend of wheat, barley, oats, alfalfa, Earthrise spirulina, chlorella, ginkgo biloba, royal jelly and 18 other herbs and super-foods. These folks even make wheatgrass pasta! Their website is the famous www.Wheatgrass.com and is the most extensive available on the internet. Even Ron Seibold's thoroughly researched book, *Cereal Grasses*, can be read there and they welcome your email questions and comments.

"We don't take out what nature puts in." —Ron Seibold

Photo Courtesy Green Foods Corp.

Harvest time on the Pacific coast of California
at the Green Foods Corporation farm

Green Magma®

Green Foods Corporation. Oxnard, CA 93030. 800-777-4430, fax 805-983-8843. http://www.greenfoods.com

Green Foods Corporation is the largest of the grass producers with divisions in both the US and Japan. Yoshihide Hagiwara, its founder, is a living legend and pioneer of barley grass *(See Pioneers)*. Although wheat grass has the name recognition, many claim that barley is the superior of these two triticum brothers. This company is a major marketing force behind the green foods revolution. It is the only grass manufacturer running advertising on national cable networks such as the Discovery Channel, Learning Channel and Lifetime and has used celebrities like golf legend Arnold Palmer to introduce green foods to the mainstream. Currently, it has the endorsement of former urologist Dr. James Balch co-author of the best seller *Prescription for Nutritional Healing.* Dr. Balch says: *"The Variety of vitamins and minerals found in barley essence is unmatched by any other single fruit or vegetable."*

More than any other company, Green Foods Corp (GFC) has put its money into research. Hagiwara, a pharmacist and researcher himself, is president of the Hagiwara Institute for Nutritional Research in Tokyo and has financed research at major Universities in the USA and Japan. While the therapeutic benefits of wheatgrass has depended primarily on clinical evidence and testimonials, Hagiwara and GFC have made substantial inroads into legitimizing the value of grass foods in the eyes of science and medicine.

Dr. Takayuki Shibamoto, Ph.D., is a professor and former chair for the department of Environmental Toxicology at the University of California-Davis. He has isolated a powerful antioxidant in young barley leaves called 2-0-GIV and claims that it helps lower cholesterol and is a better antioxidant than Vitamin C. He has published over 200 scientific papers and is the author of *A Strong Antioxidant Found in Young Green Barley Leaves.* Dr. Kazuhiko Kubota, Ph.D., is the author of *Studies on the Effects of Green Barley Juice on the Endurance and Motor Activity in Mice.* He has performed extensive research on barley leaf extract and discovered many pharmacological functions including anti-inflammation, anti-ulcer lowering of blood sugar levels and increased endurance. He is chairman of the Department of Pharmacy at the Science University of Tokyo. *(See Nutrition & Research)*

In spite of all this research, GFC makes no claims about their product in relation to specific diseases. Their stated purpose is prevention and the intent of their marketing efforts is to provide a product that is convenient to use and appeals to the mass market.

Growing, Harvesting & Processing

GFC is located near the Pacific coast on the beautiful Oxnard plains famous for being the world capital of strawberries. For eons, rain draining off the adjacent mountains has delivered its mineral rich content to the plains. Over time, lots of silt has built up there and in addition to the strawberries it has become home to turf farms raising marathon turf. Because the Pacific coast climate is so favorable, the grass can be grown year-round. Every sixty days organically raised barley leaves are harvested at the jointing stage. Two cuttings are taken from each field. Within minutes the grass is delivered to the new (1990) 52,000 square foot processing facility only minutes away. This is a massive, super-efficient, state-of-the-art plant that is truly impressive.

The grass is first washed thoroughly, rinsed and tumble dried without the use of chemicals, heat or detergents. Then the leaves are sent through a corkscrew press to gently extract the juice. The insoluble plant fiber is returned to the fields where it is mixed with natural fertilizers for compost. The juice is concentrated and ready for spray drying after it is combined with maltodextrin and brown rice. They serve as a carrier and binder for the powder. The huge computer controlled spray drier is a patented invention of Hagiwara designed to keep heat sensitive enzymes alive. It sprays the juice into a fine mist of dried powder at a temperature that never exceeds 98°F. The powder is then protected from oxidation, granulated and packaged. The plant has four quality control engineers on duty. Every lot is inspected for microbes and evaluated for nutrient content including enzymatic activity.

Juice vs. Fiber

There are two camps among the manufacturers in the grass industry—those that juice the grass and dry it into a powder and those that take the whole leaf and powder that. GFC is a juice company. While they acknowledge the value of dietary fiber, that is not the purpose of this product. Anyone who wants more fiber in their diet has numerous foods to choose from. Hagiwara created a precious vitamin concentrate that he designed to be fully assimilated, not slowed down or blocked by the various inefficiencies of people's digestive tracts. He tells us that fiber is indigestible and locks some nutrients inside it whereas juice is fully soluble and instantly assimilable even by people with weak digestion.

The Story Behind Maltodextrin

This is the most controversial subject about this company and all of its detractors readily jump on it. Unfortunately, consumers know little about the various additives used in foods and don't have time to learn more. Once they hear something negative, they just avoid it. But here's the story. Maltodextrin is a complex carbohydrate. Carbohydrates are sugars and starches from foods like grains, fiber and starchy vegetables. Complex carbohydrates take longer to enter the bloodstream, while simple sugars enter rapidly, causing the highs and lows in energy that hypoglycemics well know. Starches from grains (malt is commonly derived from barley) are complex. Thus rather than a quick shot of energy, they release energy over a long period of time. That makes Magma safe for diabetics. Hagiwara discovered that if he dissolved the long 17 chain maltodextrin with the grass juice, it would coat and protect the enzymes and nutrients from oxidation during the spray drying.

It is interesting to note that Hagiwara may actually be replicating nature here. Anyone who drinks tray–grown, Ann Wigmore style grass juice, knows that it is intensely sweet. It contains 60–70% natural sugars.

Since this is the therapeutic grass used by all the health clinics, is there some benefit to this sugar? Wheatgrass health professionals explain that it helps deliver nutrients into the bloodstream and then crystalizes in the bowel drawing out toxins. While maltodextrin does not spike blood sugar levels like the simpler sugars, it may still serve a similar role in Magma even though it makes up a smaller percentage (30%) of the powder. Brown rice (5%) is added to improve granulation during spray drying. It also enhances the flavor of the product and adds B complex vitamins.

Green Foods Corporation has a whole family of Magma products including Magma Plus, a superfoods combination product, barley grass formulas for your pets and beta carotene and wheat germ extracts. Their Veggie-Magma even includes broccoli sprout juice.

Green Kamut® Corporation

1-800-452-6884, fax 562-901-9575. Long Beach, California.
http://www.kamut.com

I *saw too many people die and suffer unnecessarily because they weren't willing or couldn't afford to make dietary changes. I wanted to make certain that when people made a decision to take their health into their own hands and control their own destiny, they would have access to products which would live up to their needs and expectations.*
—David Sandoval

This company incorporated in 1993 is the newest and smallest field--grass producer in the world. But in just a few short years and with only a small amount of money, it has established a considerable presence. It is unique in some important ways. This wheat grass is grown with Kamut®, a variety of wheat that is superior to common wheat in protein and many other nutrients. It is also the only field–grass producer that embraces the Ann Wigmore philosophy of wheatgrass and health. Green Kamut actually standardized their product to be nutritionally equal to fresh wheatgrass juice. Just about one gram (900 mg) of Green Kamut will provide the basic nutritional elements in one ounce of fresh tray–grown wheatgrass juice.

In addition to health, this company has a strong ecological ethic. It uses solar energy to electrify its processing plant including running the dehydrator that dries the grass. Entrepreneur David Sandoval, who conceived the idea of using Kamut for wheatgrass, refuses to provide samples of his product in packets like many others in the industry. *"It's ecologically false to put one gram of powder into a metal tin foil and plastic polymer layered packet and pollute the whole world just to get people to buy your product."*

Kamut is an ancient relative of modern durum wheat that originated in the fertile crescent of the Nile river thousands of years ago. The trademarked name[4] is the ancient Egyptian word for wheat whose root meaning is said to be "soul of the earth." It has only been grown in the U.S. since the 1980's. It is a non-hybrid, non crossbred seed internationally recognized as an heirloom grain. Its most famous asset is its rich buttery flavor, but you can even see the difference in its amber color and humped back kernel that is about twice the size of regular wheat. Although one might assume genetically manipulated grain would be more nutritious, the thrust of modern agricultural engineering has been higher crop yields and the improvement of flour characteristics. Non-hybrid Kamut is actually more nutritious with 17.3% protein compared to an average 12% in common wheat. Sixteen out of 18 of its amino acids rate higher than common wheat and it contains 14 chromosome pairs compared to 22 in regular wheat. Kamut is also higher in eight out of nine minerals, contains more lipids and fatty acids and has significantly more magnesium, zinc and vitamin E.

How and Where It is Grown

The high mountains and ancient volcanic hills of Utah provide the ideal setting for the land where Green Kamut is grown. Their 2, 800 acre organic farm lies in a 5,000 foot high valley surrounded by the Grand Canyon and the Rockies. Runoff from the mountains produced a lake thousands of years ago that still exists as a subterranean reservoir and provides a source of rich minerals for irrigation. The top soil extends a deep 18 feet and up to 99 feet in spots. Sandoval compares this mineral rich ancient seabed to the famous fertile crescent of the Nile. They grow "virgin grass" there, taking only one cutting at the jointing stage from each crop after approximately six weeks of growth. One crop is planted per field per year and only one harvest is made on it. Each field alternates growing alfalfa one year then Kamut the next. This loads the soil with minerals and nitrogen that increases the chlorophyll content of the Kamut. The high altitude increases its protein. Green Kamut is actually a two ingredient product made up of 65% Kamut grass and 35% mature alfalfa. Alfalfa is loaded with minerals and one of the richest sources of chlorophyll on the planet. This superb nutritional combination produces chlorophyll counts of 2–3% which is equal to chlorella algae. They also produce a 100% barley grass powder on this same soil. It's called "Just Barley" because it has nothing but pure barley grass powder in it. Both of these products are highly alkaline forming foods which is just what our bloodstream wants.

"People examine purity first and price second. If the technology exists to dry grass without additives, then a 100% pure juice extract is clearly the superior choice." —David Sandoval

Processing and Packaging

Once the grass is harvested, it is washed and ground into a pulp. The pulp is chilled and juiced with a press. The moisture is then evaporated in a unique method that never exceeds 88°F., making this a non-pasteurized product. Their technique spreads out the molecules of the grass so it can dry quickly at low temperature and is the reason the product is so light in weight. The drier is heated by the company's new solar generator. The powder is then packaged in a green tinted square PET plastic bottle with a desiccant pack to absorb moisture and sealed with a tamper proof lid. "Just Barley" is bottled in clear glass protected from light by a 100% post consumer waste box. No excipients, binders, fillers, starches or anti-caking agents are added to either product. Because the processing plant is on site, the entire transition from field to bottle takes 30 minutes.

Scientists have discovered that green juices increase the oxygenation of the body, purify the blood and organs, aid in the metabolism of nutrients and counteract acids and toxins. Green juices are the superstars of the nutrition world.

—David Sandoval, Green Kamut

Sweet Wheat™ Inc.

888-22-SWEET. Fax 813-446-5454. Clearwater, Florida.
http://www.sweetwheat.com

Started in 1997, this is the newest and smallest producer of powdered grass and unique in several ways. Compared with the other producers, this grass is different in how it is grown and dried. It looks and tastes different, too. SweetWheat, Inc. is the only bottled grass product producer in national distribution that grows in greenhouses (five of them) and in New England (their corporate headquarters are in Florida). The art of indoor tray growing of grasses was developed by Ann Wigmore and this company is a true follower of Dr. Ann. If you are a home grower, a connoisseur of juice bar wheatgrass or a student at any of the wheatgrass retreat centers, this juice will taste like the real thing. Its taste is totally unique compared with field–grown grass. This is because tray grown grasses are very high in natural grain sugars while field grasses have only 15–20% total sugars. So, this grass has that same sweet home-grown, fresh squeezed taste. And, it's certified Kosher, too.

Growing & Juicing

Growing indoors has many differences over outdoor farming and although this is not the place to discuss its pro's and con's, one of the biggest differences is the management of the soil. This company actually makes its own soil from 5 different ingredients including rock dust powder—a highly concentrated source of minerals and trace minerals. The organic wheat is sprouted in 1–2 inches of this soil and grown to a height of six inches over 14 days. Field grass is cut between 6 weeks and 200 days. Temperature and humidity are computer controlled. What farmer wouldn't want to control the weather! Only natural sunlight is used. As soon as the grass is cut, it is juiced in a high volume two stage custom juicer. First the grass blades are crushed and pulverized, then a press squeezes out all the water. The juice is flash frozen, locking its nutrients and enzymes in suspension. In a vacuum chamber, the vapor molecules are drawn off the ice leaving only crystallized flakes of juice. The enzymes and nutrients remain in suspension in these crystals until they are reactivated by adding water. The company proudly shows Kirlian photos of their product that capture on photographic paper the radiation field emanating from the product indicating the presence of a "life-force."

The flakes are then packaged in a climate controlled chamber in PET1 amber bottles. No additives, fillers, binders or preservatives are used, just 100% pure freeze dried, organic certified wheatgrass juice. Their small bottle is great for traveling and since grass is mostly water, very lightweight. All growing, juicing and packaging is done under one roof. While not in wide distribution yet, it is out there, so keep looking or contact the company through their website.

Kyo-Green

Wakunaga of America Co. Ltd. Mission Viejo, CA 92691. 800-421-2998. 714-855-2776. Fax 714-458-2764. http://www.kyolic.com

Wakunaga is a Japanese–American company originally established in Japan in 1955 with its American division located in Mission Viejo, California since 1986. This company is best known for Kyolic, the famous odorless aged garlic extract. It has a small line of botanicals demonstrating its commitment to a selected group of herbs and phytochemicals. Wakunaga is not a grower of grasses nor do they produce a 100% grass product but they are listed here because they are a longtime player in the grass foods world and have a substantial presence on the market. They also have a reputation for quality. In 1991, Mr. Wakunaga was honored with Japan's Minister of Science and Technology award.

Their Kyo-Green is a grass combination product described as the best of land and sea. Wheat and barley grasses are grown for Wakunaga in the pristine Nasu highlands of Japan. Both are harvested at the jointing stage and combined with Bulgarian chlorella. Chlorella is a single cell algae, cultivated by man, that is 2–3% chlorophyll—the richest source on the planet. (Grasses and alfalfa are a close second.) Chlorella is 55–65% protein by weight, making this a superb protein powder. The product also includes northern Pacific grown kelp and brown rice. The rice serves as a carbohydrate and fiber necessary for the spray drying of the juice. Kyo-Green melts readily in water or juice and has a smooth taste. In addition to the full spectrum of nutrients found in grasses, this blend is an especially good source of vitamin B-12, protein and iodine.

VitaRich Foods

Naples, FL 800-817-9999, fax 941-591-8220. http://www.VitaRich.net

Vita-Rich started in 1992 and although fairly new on the scene, it has become a major player. This company is unique in that it does not produce its own brand. Instead it is the manufacturer behind the scenes of several popular grass and green food mixtures in the U.S., England, Australia and South Africa. Vita-Rich's main products are barley and wheat, but it also cultivates a macro algae called Hydrillia. Macro means this algae develops roots. Hydrilla is even richer in minerals than spirulina and chlorella.

Although headquartered in beautiful Naples, Florida, its farm operation lies on the Northern Florida–Georgia border near Tallahassee. This provides a perfectly cool climate for wheat and barley. Their 2,500 acre farm has a high clay content that is ideal for mineral and moisture retention. Fertilizers stay on top longer, feeding the developing roots rather than percolating down beyond their reach. The fields are irrigated from an 800 foot deep well that delivers a rich mineral water. Three hundred acres of grain are planted and harvested from November through April. A laser sighting system helps level the land or angle it slightly for drainage. This smooth surface enables the harvester to get really close to the ground. The grass is cut just before jointing stage when the blades are about 8–9 inches tall. The young, sweet blades are soft and luscious and free of hard stalk fiber. When the stalk starts to develop, it creates a 'seed head' which drains the nutrients from the blades. This was the discovery of Charles Schnabel *(see Pioneers)*. Vita-Rich is able to get two cuttings per location before jointing.

Processing the Grass

Harvesters carefully convey the grass to the processing plant that is centrally located on the acreage. There the grass is put through a five tank washing system that includes ozonation to control microbe activity. Then the grass is dried and ground.

Their low temperature 89°F drying system consists of a six foot wide conveyer that rides the grass leaves through a high volume dehumidifier over and over again until all the water is evaporated. It can take over four hours to dry. This is longer than any other manufacturer because Vita-Rich president Kevin Thomas insists on drying the whole, intact leaf.

"The more we crush, pulverize or even bruise the leaves, the more we lose enzymatic activity. Drying the leaf whole keeps it fresher. It even smells better this way."

Once the leaf is dry and brittle, it can be powdered with less risk to its nutrition. Vita-Rich sucks the grass through a milling head that is kept to a cool 50°F with nitrogen. This protects against the higher temperatures generated by friction. The grass blades bounce and spin around until they become small enough to travel through the filter. By this time it has become a micro-fine particle that easily dissolves.

No Juice But You Won't Notice

Vita-Rich used to make grass juice using a spray dryer system like that developed by Hagiwara. But Thomas did not like all the mastication and pulverization of the plant along with the starch additives necessary as a carrier. He found that just grinding up the whole leaf was simpler, healthier and less expensive for the consumer. Their grinding method creates such fine particles, they dissolve nearly the same as a juice powder. Since grass is mostly water anyway, getting the few solids out creates more processing, more expense and more risk to the nutritional integrity of the plant than it's worth. Pines of Kansas also takes this approach.

> *"This is a juice and fiber concentrate. It has 100% of juices and solid fiber minerals, enzymes, and proteins that are attached to the plant fiber cells. It dissolves readily because it is micro-pulverized. And it hasn't been diluted by the additives necessary for making juice powders."* —Kevin Thomas, Vita-Rich

Since Vita-Rich does not produce its own brand, safe transportation to the various manufacturers is required. The dried grass powder is shipped by Styrofoam insulated air freight containers and arrives the next day even in places as far away as South Africa. By land, it travels in insulated, air conditioned trucks.

Vita-Rich does one more thing that is very unique. It uses strong magnets during the grinding stage to align the atoms of grass in a uniform polarity. Although this is an esoteric goal, it is based on immutable laws of physics. The concept is that properly polarized food will be handled more efficiently by the body, achieving more complete combustion and greater absorption. Since grass customers are somewhat esoteric to begin with, this should not be a stretch for them. The magnets also have a practical application in that they eliminate static electricity built up from processing. Magnets are already popular in the health movement therapeutically for relief of pain, on home water systems for treatment of hard water, on auto fuel lines for improving gas mileage and in acupuncture treatment.

Greens+®

Orange Peel Enterprises. Vero Beach, FL 32960. 800-643-1210
fax 561-562-9848. http://www.greensplus.com

Greens+® is the original grass and superfoods combination product. Looking at the dozens of green super-food powders on the market today, it is hard to believe that in 1992 there weren't any. But that is the year that Sam Graci premiered Greens+. Sam, author of *The Power of Superfoods*, is a chemist and psychologist by profession. In the 1970's, he worked with a group of Down Syndrome teens attempting to improve their social interactive skills. In the process, he discovered they had extreme vitamin, mineral and enzyme deficiencies. Upon improving their diet and adding rest and exercise, he saw dramatic changes. Sam continued to do nutritional research with help from some very important experts including Dr. Linus Pauling. He wanted something that would be easier for people to take than dozens of bottles of different colored vitamins. The result was an easy-to-mix powdered blend of super-foods. Sam believed that concentrated whole foods with their full spectrum of nutritional cofactors were more effective than isolated vitamins. The result was a synergistic blend of enzymatically active, alkaline–forming, nutrient-rich foods in an easy-to-take powder. At its foundation are the grasses of barley and wheat along with alfalfa and the algaes: spirulina, chlorella, dulse and dunaliella. On top of this, he added organically grown soy sprouts, antioxidant herbs and extracts, probiotic cultures and pectins, royal jelly, ginseng and all in all, about 29 different herbs and super-foods. He called this rich chlorophyll mixture Greens+® and arranged for Orange Peel Enterprises to manufacture it. Today, every major nutritional supplement manufacturer has its own version. Sam pioneered a new niche in the nutritional foods industry and barley and wheat grasses are at the foundation of every one of them.

Although they have much competition, Greens+ is still the industry leader. In 1996 they received the International Hall of Fame New Product award in the health category and took first place in the annual People's Choice Awards bestowed by the National Nutritional Foods Association. According to the Boston Herald, even David Letterman uses Greens+ and swears by it. These premium green foods raise energy levels, build mental acuity, strengthen the immune system, maintain colon ecology, balance pH, and enhance metabolism.

What follows is a listing of a few of the many manufacturers of powdered green super-food concentrates. Unlike the primary field grass producers who grow and sell under their own brands, these manufacturers purchase their grasses from the previously described "producers." Greens+® grasses, for example, are grown by Vita-Rich. Keep in mind that because of the emerging popularity of green foods, most large supplement manufacturers now provide a green foods powder product. So check your favorite brand and ask at your health food store for others. For additional listings, check the world wide website for this book at: http://www.Sproutman.com .

NutraGreens by Earthrise. 800-949-RISE, 707-778-9078, fax 707-778-9028. http://www.spirulina.com http://www.earthrise.com

Green Magic. New Spirit Naturals. 888-337-4657 or 407-805-0510 Fax (407) 333-3286. http://www.newspiritnaturals.com/

Miracle Greens. Fit For You Int'l, Inc. Brentwood, CA. 800-521-5867. Fax 310-207-2638.

Greens Today. The Organic Frog. 800-GREEN-41. Fax 516-921-6537. http://www.organicfrog.com

Science & Wheatgrass
The Quest for Confirmation

Thomas Alva Edison, after working for 60 straight hours perfecting the phonograph. As the inventor of the phonograph, motion picture, electric light and telephone enhancements, this man laid the foundation for the twentieth century.

Until Man duplicates a blade of grass, nature can laugh at his so called scientific knowledge. —Thomas Alva Edison

What role does science play in confirming wheatgrass as a natural medicine? In this book there are numerous research citations demonstrating its wondrous healing properties. Every decade since the 1930's has produced studies on its prodigious nutritional and phytochemical qualities. The magical properties of chlorophyll as a blood builder for the anemic, for disinfecting and neutralizing odors, and healing wounds started becoming known as far back as 1915. The 1970's, 80's and 90's presented numerous studies on barley grass and its superior antioxidant, tumor suppressant and immune supportive functions. These were traditionally designed studies with many published in peer-reviewed journals. Nevertheless, grasses are not a mainstream accepted food or medicine and their therapeutic potential benefits only a closed club of faithfuls. It is considered unproven. But clinical evidence from health professionals and testimony from users who have conquered life threatening diseases after being abandoned by conventional medicine have value. Do we need to wait until the therapeutic benefits of grass are scientifically proven before we use it?

Gray are all the theories, but Green is the Tree of Life. —Goethe

What Do You Believe?

Is all that is true only so because science has pronounced it? How are we to evaluate the "truth" without science? The extent of information we are willing to accept without scientific proof depends on our belief system.

Divine Belief. If you have a divine or intuitive belief system, then you trust in God and nature. You are likely to ask: "Do we have to understand why everything works before we try it?" *Faith* underlies the foundation of this belief system. "And the truth shall bear witness of itself." "To feel beauty is a better thing than to understand how we come to feel it."—Santayana[1]

Agnostic–Darwinian System. If you have an agnostic–Darwinian belief system, you require proof. You depend on science and are skeptical of anything unproven. "There is something Pagan in me that I cannot shake off. In short, I deny nothing, but doubt everything." —Lord Byron, English poet (1811). "No scientist, however specialized his field, can factually accept even the Book of Genesis."—Robert Graves, (1963) British novelist.

What greater means of substantiation is better than scientific proof? Society does need a standard to measure things against. But science is a big world. Scientists themselves are frequently the first to look askance at the studies of other scientists. As a group, they are the ultimate skeptics. Even when proper protocol is followed, many still proclaim disbelief. If doctors and scientists dispute the results of studies, what are we as consumers supposed to believe? Throughout medical history, studies have contradicted one another and previously accepted positions have been reversed. In two recent studies on calcium, the title of the first read: *Calcium Supplementation Ineffective in Replacing Bone.* The second study, performed by a different group of scientists, but using the same protocol: the same diet, same number of people, same distribution of men and women, same ages, pronounced: *Calcium Effective in Replacing Bone.* Which one are we supposed to believe? It turns out that the only aspect of the two studies that was not identical was the source of the calcium. The first group used calcium carbonate, the second used calcium citrate. The first group insisted: "calcium is calcium." But calcium carbonate is not readily absorbed by the body.[2]

If the doctors and scientists don't agree, what are we consumers supposed to believe?

Scientists still are not in agreement over the greenhouse effect. Yet according to the British Meteorological Service, 1995 was the warmest year in history. A United Nations panel of scientists predicts that if emissions are not reduced, climatic changes will become more and more irreversible. Are we to abide by the axiom: If you can't prove it, it's not happening? How sure do we have to be before changing the way we live? Dr. Charles F. Schnabel, a chemist to whom this book is dedicated, said upon discovering the phenomenal fertility of chickens eating wheatgrass: *I shall never forget the day in July, 1930 when I gathered 126 eggs form 106 hens. My first thought was—'How long will it be before science can explain it?'* [3]

The art of healing comes from nature, not the physician. Therefore the physician must start from nature with an open mind.
—Paracelsus 1493-1541

Where Does Medicine Come From, Anyway?

It wasn't until the modern drug industry arose in the 19th century that medicines came from the laboratory. Nature is the original source of modern medicine. Pharmacopoeias of ancient Egypt, Babylonia, Greece, and China were based on food. Hippocrates, the father of modern medicine, proclaimed, "Let your food be your medicine and let your medicine be your food." The 12th century Jewish physician/philosopher Maimonides recommended chicken soup as a remedy for asthma. Garlic, mustard seed, and other herbs and spices were used medicinally for centuries. And what child doesn't know: "an apple a day keeps the doctor away." Many modern pharmaceuticals are actually synthetic replications of botanical products. Quinine from the cinchona tree was the only remedy for malaria for over 300 years. Penicillin, our first antibiotic, comes from mold. Aspirin is synthetic salicylic acid derived from the bark of the willow tree. In spite of our high technology, researchers today still look to nature for ideas and synthesize natural compounds. If nature is the source of all these miraculous medicines, why has it not earned our confidence? Why, as a society, can't we accept wheatgrass with the current level of proof? Even scientists believe in things. They thrive on hunches. The only difference is they substantiate their beliefs in the lab while we embrace them with our "faith."

Science without religion is lame.
Religion without science is blind.—Albert Einstein, 1934

The ordinary doctor is interested mostly in the study of disease. The nature doctor is interested mostly in the study of health.
—Mahatma Gandhi, 1869-1948.

Why Your Doctor Won't Tell You About Wheatgrass

We don't do very well with our differences in America. If one thing is good, the other has got to be bad. As a profession, medical doctors are automatic doubters. They cannot give wheatgrass credit, for to accept the existence of a significant medical force outside of their domain is to diminish their power. Every group has a natural instinct to protect itself. Is the doctors' position on wheatgrass an evaluation or a reaction? They must take an extreme position in the national arena. Their rhetoric is all black and white: Alternative medicine is filled with quacks and established medicine is the only valid system. The psychology is understandable but by taking this, position conventional medicine is protecting itself rather than protecting the public. What is this if not an intolerance for a different opinion? There are Nazi hate groups today that deny the holocaust. Does that mean it never happened? Is wheatgrass a fraud because the medical profession refutes its value?

It is time we prove to America we are a doctor's organization, working for the good of our patients, rather than a pressure group aiming for political power as a way to build organizational predominance, to create personal prestige, or to line our own pockets. —former AMA President, John J. Ring, MD.

Why There Aren't More Studies on Wheatgrass

Wheatgrass is not a drug. Natural products are not easily patented and the average cost of bringing a new product up to FDA approval is $359 million per therapeutic use. While this may be only a few days' earnings in the prescription drug industry, it is beyond the reach of natural products manufacturers. Getting a study published in medical journals is difficult, and drug companies have substantial political and financial influence with them, since journals depend on drug ads to survive. Unless the system changes, it is unlikely that approved medicines will come from anyone other than major pharmaceutical companies.

Any cure for cancer or anything else from outside the system must be suppressed to maintain the status quo.

Conventional medicine takes years to accept any discovery that challenges cherished beliefs. Typically, a discovery that contradicts mainstream authority is regarded as quackery. Pasteur was reviled for years about his "germ theory." William Harvey (1628) was ridiculed when he claimed that blood circulates. Roentgen was laughed at in 1895 upon discovering X-rays. It is always an uphill fight opposing the status-quo which by its nature is against change.

> *Animals haven't heard anything about vitamins. They determine the nutritive value with their instinct, palate and olfactory faculties acting for them in place of judgement. The preferences shown by cattle are therefore better proofs than those obtained from the analysis of the chemist.*
>
> —George Sinclair, 1869 [4]

Approved Drugs Are Not So Safe

What critics often ignore is that not everything in conventional medicine is sanctioned by published science. Aspirin and penicillin became widely used long before experts knew how they worked. Many mainstream techniques such as surgery, anaesthesia and drugs are acknowledged as unsafe. In the April 1998 issue of the prestigious *Journal of the American Medical Association,* researchers calculated that adverse drug reactions may kill more than 100,000 Americans each year. In 1994, deaths from drug reactions were so numerous that statistically they placed between the fourth and sixth leading cause of death in the United States.[5] What value are published studies and expensive FDA approval if this is the result? The FDA claims the "benefits of drugs outweigh the risks." Since wheatgrass has no known toxic side effects, it has no risks, just benefits. Yet, in spite of the rather deadly track record of approved drugs, wheatgrass is still rejected as untested and unproven.

Not All Studies are Valid

Every year FDA approved drugs are pulled off the market because of incidents involving dangerous side effects, including deaths. How valuable then, is government approval? How reliable are scientific studies if, as in the case of two calcium studies, we cannot depend on the results? Can you interpret a scientific study? Since for most of us the answer to this is "no," then why do we place so much value in them? To protect ourselves, we must ask some questions.

How To Study A Study

What are the credentials of the scientists? Is it a double blind study where neither the patient nor the scientist know the source of the sample? Are placebos used? Has the study been published in a peer reviewed journal? Are the results statistically significant? Were the tests done in vivo or in vitro—in animals or in test tubes? On mice or people? Who funded it? Is there any group who will benefit from this financially? Are testimonials by doctors paid for? Look for a conflict of interest such as the cigarette industry financing a study on lung cancer. Beware of health claims that tout one product as a panacea. Health is never that simple.

Healing is Electrical, Spiritual, Chemical, Physical

Americans love pills. Modern medicine has provided us with some powerful effects through chemistry. The scientists and pharmacologists should be applauded for the powerful medicines they have developed. But healing is more than chemistry. Taking vitamins all day will not alone solve most chronic conditions. The restoration of health is complex. There is no magic pill. Wheatgrass is neither a pill nor a panacea. But when used as part of a total health revitalization program with an indefatigable commitment to wellness, it is a powerful healing agent that will maximize and accelerate your health potential.

Health is more than nutrition—more than the physical feeding of chemicals to cells. We are more than just physical, mechanical bodies. We are bio-electrical, spiritual beings inside physical bodies. One of the secrets to the effectiveness of wheatgrass is that it nourishes us on these levels, too. As a high vibration living food, it raises our chi to a higher vibratory level. In the drug culture's vernacular, you get high. It's the difference between walking and jogging. Wheatgrass accelerates the electrical activity inside every cell. That doesn't happen with canned carrots, powdered potatoes, french fries or microwave pizza. Chocolate makes you feel good because it temporarily spikes your blood sugar. Coffee is a chemical stimulant. But grass enlivens you. It's like charging your biological battery. It's a natural high, boosting your own manufacture of adrenalin. It's how you feel when you're on top of the world. You can conquer any challenge, defeat any foe. This is the spiritual factor that is a necessary part of any successful battle with illness. The Chinese call it chi; the yogi's call it kundalini—the God–force, the life force. With it behind you, you can go far and you can go fast. You are the healer. Wheatgrass is the energizer.

The Final Analysis

Although many people have heard about wheatgrass, it remains largely unknown and undiscovered in the general population. When you mention “grass” to the masses, they think you mean marijuana. But for the thousands who are turned away each day by conventional medicine, wheatgrass is hope, not dope. In the world of alternative medicine, grass is king. Whether or not you believe in it, it has earned the right to be evaluated by more people—before they enter a desperate health crisis. This book provides the tools to help you design your own alternative health rejuvenation program.

Wheatgrass is hope, not dope

Do you need scientific verification? That is a question you must answer for yourself. But a successful health journey must include faith. If you were to guess whether wheatgrass will benefit your health, what would your answer be? Take charge of the direction of your personal health journey. It is an illusion to believe that doctors, hospitals, scientists or anyone else knows better than you. You already have the knowledge to make the right decision. Now you need only muster the courage.

Sproutman

Let’s change “Impossible” to “I’m Possible.”
—Dr. Bernard Jensen

If I have the belief that I can do it, I shall surely acquire the capacity to do it even if I may not have it at the beginning.
—Mahatma Gandhi

Epilogue

Photo by Michael Parman

by Dr. Ann Wigmore

Today in the twentieth century, we must learn to survive, not only from the hazards of pollution in our water and air, but also from our food supply, itself. Many Americans are slowly killing themselves by what they eat each day. Thousands of chemical additives havc found their way into our daily foods and the foods themselves are overcooked and overrefined during processing, storage, delivery and marketing. Staples like bread and cereals are so depleted of their own naturally occurring vitamins, minerals and bran by over-processing and refining that laws have been passed forcing manufacturers to fortify them with synthetic ones.

Malnutrition amongst the world's affluent peoples is becoming more and more common. It is the natural result of eating too many foods which have been processed and cooked to the extent that they are depleted of their naturally occurring nutrients. However, we have a choice of how we want to live and what we want to put into our bodies.

For almost thirty years I have been teaching people all over the world how to grow their own highly nutritious live foods like sprouts, indoor garden greens and wheatgrass at less than half the cost of supermarket foods. I have also lived almost entirely on these foods which I grow from seed in my home in Boston and even while I am traveling.

It is easy to control your own food supply and take responsibility for one's health. Time has shown that our battles against degenerative and chronic diseases like cancer, heart disease, diabetes, arthritis and many others cannot be won by palliative measures like surgery, chemotherapy and other medical intervention, but only through prevention. Unfortunately, more people wait until there is a critical problem to give preventative action a thought and then it becomes a question of preventing a recurrence. But true prevention stops a problem before there is any serious permanent damage. The fact is survival is not only possible but joyous. It is the re-inheritance of our natural birthright–health–which has been stripped from us by modern man's errors in judgement.

—*Ann Wigmore,* Boston, 1984.

Ann Wigmore died in Boston on February 16, 1994, in the Mansion where she taught, of smoke inhalation from an electrical fire. She was 84.

The Ann Wigmore Foundation is now located at PO Box 399, San Fidel, NM 87049. 505-552-0595. Her retreat center in Puerto Rico is the Ann Wigmore Institute, PO Box 429, Rincon, PR 00743, 787-868-6307.

References and Studies

HISTORY AND CULTURE

1. Biogenic Meditation Biogenic Self-Analysis Creative Micro-Cosmos. Students' and Teachers' Digest and Guidebook to Intensive Biogenic Seminars. International Biogenic Society. 1978. See Resources: I.B.S. and Awareness Institute.

2. *Grasses An Identification Guide,* by Lauren Brown. Houghton Mifflin, 1979.

3. Ibid

4. *Hortus Gramineus Woburnensis* or, an Account of the Results of Experiments on the Produce and Nutritive Qualities of Different Grasses, used as the Food of the More Valuable Domestic Animals. By George Sinclair, Gardener to the late Duke of Bedford. Instituted by John Duke of Bedford. Published by Ridgways. London, 1869. 362 pages.

5. Kellner, Jordan, Wilson, et al. Missouri, 1915. From The Cause and Cure of Famine and Hidden Hunger, by Charles F. Schnabel. 1936. Unpublished.

6. Second Annual Report of the Department of Agriculture state of New Hampshire, 1890 by Morse, F.W.

7. Woodman, H.E. & associates. *Journal of Agricultural Science,* 1925–1930.

8. *Hortus Gramineus Woburnensis.* (See previous note) By George Sinclair. London, 1869.

SPIRITUAL AND RELIGIOUS ROOTS

1. Song of Myself by Walt Whitman, 1819-1892, sct. 31, in *Leaves of Grass* (1855).

2. *George Washington Carver–An American Biography* by Rackham Holt. Doubleday, Doran & Co., 1943.

3. *The Meaning of the Word Grass in Hebrew.* Published by The Western Wheatgrass Journal, VII, No. 1. Jan-Mar. 1995. 3353 South Main, Salt Lake City, UT 84115. See *Resources.*

4. *Written By the Finger of God—Decoding Ancient Languages,* by Joe Sampson, Oct. 1993. Wellspring publishing, Box 1113 Sandy UT 4091. ISBN 1884312-05-5.

5. Op.cit.

6. See chapter: *Nutrition*

7. Conversation with Piter U. Caizer, by Steve Meyerowitz, California, March 15, 1998. Edited by Steve Meyerowitz. All rights reserved. Permission for use in any medium is required.

8. Ibid.

9. *The Essene Gospel of Peace Book IV*, by Edmond Bordeaux Szekely. Chapter: "The Gift of the Humble Grass." International Biogenic Society. (Address below.) Used with permission.

10. The Awareness Institute. 1305 S. Highland Park Drive, Lake Wales, Florida 33853-7471. Phone 941/676-6231. http://www.awarinst.com/

11. *International Biogenic Society.* P.O. Box 849, Nelson, B.C. Canada V1L 6A5.

12. *The Essene Gospel of Peace Book IV*, by Edmond Bordeaux Szekely. Chapter: "The Gift of the Humble Grass." International Biogenic Society. Used with permission.

13. Extracted from *"Grass is the Forgiveness of Nature."* The author was the senator from Kansas in the late 1890's.

THE PIONEERS

1. *Letter to Conrad A. Elvehjem*, July 10, 1942. Personal papers.

2. *"The Place of 40% Protein Grass in Modern Agriculture,"* by C.F. Schnabel. Unpublished papers. 1939.

3. *"Good Grass is the Basis of A Permanent Agriculture,"* by C.F. Schnabel.

4. "Peace from Grass" by C.F. Schnabel, B.S., D.Sc. *Cappers Farmer,* October, 1947.

5. JAMA. *Journal of the American Medial Association.* "Council on Foods: Accepted Foods." Franklin C. Bing, Secretary 1939. 112:733.

6. *The Relation of the Grass Factor to Guinea Pig Nutrition.* By G.O. Kohler, C.A. Elvehjem, and E.B. Hart, Department of Agriculture Chemistry, University of Wisconsin, Madison. Journal of Nutrition, Vol.15, No.5. 11/24/1937

7. "The Amazing Anti-Aging Diet," by Claudia D. Bowe, *Cosmopolitan,* November, 1983.

8. *The Hippocrates Diet and Health Program*, by Ann Wigmore. Avery Publishing Group, Wayne, NJ. 1984. Dennis Weaver, actor from forward, regarding Dr. Ann's book.

9. *How I conquered Cancer Naturally,* by Eydie Mae Hunsberger and Chris Loeffler. Harvest House Publishers.

10. *The Wheatgrass Book,* by Ann Wigmore. Avery Publishing Group.

11. *Young Barley Plant Juice,* by Yoshihide Hagiwara, M.D. pub. by The Green and Health Association, Tokyo, Japan April, 1980.

12. *Ann Wigmore Foundation.* PO Box 399, San Fidel, NM 87049. 505-552-0595.

13. *Green Barley Essence: The Ideal Fast Food*, by Yoshihide Hagiwara, New Canaan, CT. Keats Publishing, Inc. 1986.

14. *Young Barley Plant Juice,* by Yoshihide Hagiwara, M.D. pub. by The Green and Health Association, Tokyo, Japan April, 1980.

REFERENCES AND STUDIES

NUTRITION

1. Dr. Charles F. Schnabel, father of wheatgrass in a letter to scientist Conrad A. Elvehjem, discoverer of niacin, July 10, 1942.

2. *The Cause and Cure of Famine and Hidden Hunger.* 1936. Charles F. Schnabel. Unpublished.

3. Phillips & Goss. Journal Agricultural Research. Vol. 51:p. 301. 1935.

4. *Chlorophyll. Nature's Green Magic* by Theodore M. Rudolph, Ph.D. Nutritional Research Publishing Company. PO Box 489, San Gabriel, CA. 1957.

5. Food Science for All, New Sunlight Theory of Nutrition. E. Bircher, Health Research Press.

6. Chlorophyll and Hemoglobin Regeneration After Hemmorrhage, by J.H. Hughes and A.L. Latner. Journal of Physiology. Vol.86, #388, 1936 University of Liverpool.

7. Chlorophyll, Nature's Green Magic, Dr. Theodore Rudolph.

8. The Influence of Diet on the Biological Effects Produced by Whole Body Irradiation. M. Lourou, O. Lartigue, Experlentai, 6:25, 1950.

9. Further Studies on Reduction of X-irradiation of Guinea Pigs by Plant Materials. Quartermaster Food and Container Institute for the Armed Forces Report. N.R. 12-61. by D.H. Colloway, W.K. Calhoun, & A.H. Munson. 1961.

HEALING WITH GRASS

1. The Wheatgrass Book by Dr. Ann Wigmore, Avery Publishing Group, Garden City Park, NY.

2. Performed at Hippocrates Health Institute in the 1970's by Ann Wigmore and staff.

3. See: *Juice Fasting and Detoxification* by Steve Meyerowitz. 1996. ISBN#1-878736-64-7.

4. The Place of 40% Protein Grass in Modern Agriculture, by C.F. Schnabel. Unpublished. 1939.

5. Numerous cookbooks are available for juice recipes including "Sproutman's Kitchen Garden Cookbook," by Steve Meyerowitz. ISBN# 1-878736-84-1. 1994. 336 pgs. ppbk.

6. Journal of Personality and Social Psychology, June 1998. By S.C. Segerstrom, University of Kentucky in Lexington, Psychology dept.

7. Science Newsletter, 1941 by Dr. R. Redpath and T.C. Davis.

8. Chlorophyll: An Experimental Study of its Water-Soluble Derivatives in Wound Healing. by L.W. Smith and A.E. Livingston. American Journal of Surgery. 52:358, 1943.

REAL STORIES FROM REAL PEOPLE

1. Courtesy of *The Wheatgrass & Sprouting Journal.* 3353 South Main, Suite 197, Salt Lake City, UT 84115. See *Resources.*

2. Mr. Lampro was a student at *Hippocrates Health Institute* in West Palm Beach, Florida *(see Retreats)* where he established his wheatgrass and health program.

3. Russell Rosen, director Naples Institute of Optimum Health & Healing. See Retreats.

4. Julian Whitaker, MD is editor of the *Health and Healing Newsletter*, the largest natural health newsletter in the world. He is also author of *New Cures for Chronic Diseases*, which is available free to newsletter subscribers. He operates the *Whitaker Wellness Institute* in Newport Beach, CA. To receive the *Health and Healing Newsletter* call 800-539-8219. To find out more about the *Whitaker Wellness Institute*, call 800-826-1550.

5. The Place of 40% Protein Grass in Modern Agriculture by C.F. Schnabel, Ph.D. Unpublished papers. 1939.

6. Philippus Paracelsus, 1493-1541, was a German born physician who introduced the concept of disease to medicine.

HEALING RESORTS

1.The Hippocrates Diet and Health Program by Ann Wigmore. Avery Publishing Group, 1984.

2. Twenty cents per minute rate to Australia from the USA in 1998 was available through Tel-Affinity Corportaion. 800-338-3202, or 781-433-0451, fax 781-433-0951.

3. This is also the site of the annual 'Arkansas Womyn Festival' where numerous performing artists for dance, music and other creative arts convene.

GROWING YOUR OWN GRASS

1. See Resources–*Mail Order Growers and Shippers*, and *Seed Sources* for vendors that sell automatic sprouters and wheatgrass growers.

2. The author-designed sprouter is called "The Sprout House Table Top Greenhouse." This product is wholly owned and distributed by The Sprout House, an independent California company owned by Rick Kohn. The Sprout House sells Sproutman designed products and approved seeds. To contact them, call 1-800-S-P-R-O-U-T-S. For more information about them and other such vendors, see the Resources chapter.

THE COMPANIES

1. Even the space taken to discuss these companies should not be construed as an indirect recommendation. It is more likely related to the history and longevity of the company and the depth of the materials available about them.

2. Consumer testimonial courtesy of Pines International, Inc.

3. Prices are suggested retail as of 1998. Street prices may be lower.

4. Kamut is a registered trademark of Kamut International Ltd. of Fort Benton, Montana and is listed with the Department of Agriculture as QK77. http://www.kamut.com

SCIENCE AND WHEATGRASS

1. *The Sense of Beauty,* by Spanish-born, American philosopher George Santayana, 1896.

REFERENCES AND STUDIES

2. *Glutathione and Its Many Important Co-factors*, by Alan Pressman, D.C., Ph.D. Fourth Annual Organic Living Conference, 1997, New York State Natural Foods Association.

3. *Letter to Conrad A. Elvehjem, July 10, 1942.* Unpublished papers.

4. Hortus Gramineus Woburnensis or, an Account of the Results of Experiments on the Produce and Nutritive Qualities of Different Grasses, used as the Food of the More Valuable Domestic Animals. By George Sinclair, Gardener to the late Duke of Bedford. Instituted by John Duke of Bedford. Published by Ridgways. London, 1869. 362 pages.

5. Journal of the American Medical Association. April 1998 issue.

RESOURCES

Where to Go to Get More Information

Mail Order Grass Growers & Shippers

A partial list of grass growers and mail order shippers. For additional up to the minute listings, see "wheatgrass" at http://www.Sproutman.com

The Sprout House. 1-800-S-P-R-O-U-T-S. 760-788-4800, fax 760-788-7979. Website: http://www.SproutHouse.com/ E-mail: info@SproutHouse.com. Fresh organic wheatgrass and sprouts shipped overnight. Dry grass powders. Sproutman tested organic sprouting seeds plus wheat, barley and Kamut grass seeds. Hydroponic automatic wheatgrass growers. Sproutman designed sprouting kits. Sprout chart and books. Ships nationwide overnight. Rick Kohn, owner.

Gourmet Greens. 802-875-3820. Dodge Road, Chester, VT 05143. E-mail: greens@gourmetgreens.com. Next day mail order certified organic wheatgrass and sprouts plus seeds and organic soil to grow your own. New and used wheatgrass and vegetable juicers. Richard Rommer proprietor, grew grass for Ann Wigmore at the original *Hippocrates Health Institute*. Website: http://www.gourmetgreens.com

Greensward Nurseries & New Natives Sprouts. 408-728-4136, fax 408-761-3956. PO Box 1413, Freedom, CA 95019. Certified organic growers of wheatgrass and 12 varieties of sprouts. Ships nationwide.

Wheatgrass Express. Certified organic grass delivered fresh overnight to your door. 15117 N. State Rd. 121, Gainesville, FL 32653. 1-800-859-4779. Fax 904-462-7398. http://www.wheat-grass.com

Perfect Foods. 800-933-3288. 4 Hawks Nest Road, Monroe, NY 10950. Frozen grass, juicers, trays, seeds and soil. The oldest and largest wheatgrass supplier in New York. Harley Matsil, proprietor.

Green Glen Farms. 707-833-6647, fax 707-935-0662. Richard Kersh is a quality northern California organic wheatgrass and sprouts grower. Whtgrass@pacbell.net Box 418 Kenwood, CA 95452. Will ship grass.

SuperFood Provisions. Tablets and powder of barley and wheat grasses, spirulina and chlorella via mail order. 800-544-3657.

Some Regional Grass and Sprout Growers

Jonathan's Sprouts. 508-763-2577. PO Box 128, Marion, MA 02738. New England. http://www.panix.com/~sprouts

Snider's Sprouts. 301-424-7878. 12341 Glen Mill Road, Potomac, Md. 20854. Southeast region–Maryland, Wash. DC and Northern VA.

Kowalke Family Sprouts. 310-455-1901, fax 310-455-3121. 1956 Old Topanga Canyon Road, Topanga, CA 90290. Southern California.

Real Live Foods. PO Box 937, Niwot, CO. Large grass grower in the Boulder, Colorado area. 303-652-2008, fax 303-652-8858.

Lifeforce Growers. 781-894-3183. Fax 781-893-0363. 46 Howard St., Waltham, MA 02154. Organic wheatgrass and sprouts. Boston area.

Sproutime. 3621 Mountain View Ave, Los Angeles, CA 90066. Tel. 310-397-6119, fax 310-390-5281.

Sunflower Farms. 12033 Woodinville Dr. #22, Bothell, WA 98011. Phone 206-488-5652. Wheatgrass, sunflower, buckwheat sprouts.

Pekula Sprouts. Mark Solman. Tel. (760) 9-4-GRASS. 957 Urania Ave Leucadia, CA 92024.

Grass Roots Organics. Deborah Klaven. 925-443-9223. PO Box 3756 Livermore, CA 94551. Large grower in Bay area.

The Sproutman of Pennsylvania. PO Box 308, Upper Black Eddy, PA 18972. 610-982-9108, fax 610-982-9107. Quality grass and sprouts.

Growers and Manufacturers

Pines International, Inc. PO Box 1107, Lawrence, KS 66044. 1-800-MY-PINES, (800-697-4637) Fax: 913-841-1252. Website: http://www.wheatgrass.com E-mail: InfoService@Wheatgrass.com

Green Foods Corporation 320 North Graves Ave, Oxnard, CA 93030. 800-777-4430, fax 805-983-8843. http://www.greenfoods.com gfc@greenfoods.com. Makers of *Green Magma* barley grass juice.

Green Kamut Corporation. 1542 Seabright Ave., Long Beach, CA 90813. 1-800-452-6884, fax 562-901-9575. http://www.kamut.com

Sonne, V.E. Irons. Makers of *GreenLife*. Office: PO Box 2160, Cottonwood, CA 96022. 800-544-8147, fax 916-347-5921.

SweetWheat Inc. PO Box 58, Chaplin, CT 06235-0058. 1-888-227-9338. Freeze dried grass juice. http://www.sweetwheat.com

VitaRich Foods. 800-817-9999. http://www.VitaRich.net/ fax 941-591-8220. Naples, FL 34109. Wholesale inquiries only.

Kyo-Green. Wakunaga of America Co. Ltd. 23501 Madero, Mission Viegjo, CA 92691. 800-421-2998. 714-855-2776. Fax 714-458-2764. http://www.kyolic.com/ Japanese grown grass and chlorella.

Greens+. Orange Peel Enterprises, Inc. 2183 Ponce de Leon Circle, Vero Beach, FL32960. 800-643-1210, fax 561-562-9848. http://www.greensplus.com/ First multi-green powder product.

Healing Resorts that Use Grass

Naples Institute for Optimum Health & Healing. 800-243-1148. 941-649-7551. Fax 941-262-4684. http://www.NaplesInstitute.com/ 2335 Tamiami Trail N. Naples, FL 34103.

Optimum Health Institute - San Diego. 800-993-4325. Tel. 619-464-3346, Fax 619-589-4098. http://www.optimumhealth.org/ 6970 Central Avenue, Lemon Grove, CA 91945-2198.

Optimum Health Institute - Austin, Texas. 800-993-4325, Reservations. Tel. 512-303-4817, Fax: 512-332-0106. Rt. 1, Box 339-J Cedar Lane, Cedar Creek, TX 78612

Ann Wigmore Institute - Puerto Rico. PO Box 429, Rincon, PR 00677 USA. Tel 787-868-6307, Fax 787-868-2430. Wigmore@caribe.net

Ann Wigmore Foundation. PO Box 399, San Fidel, NM 87049. 505-552-0595, fax 505-552-0595. http://www.wigmore.org/~wigmore/ Fifty-eight miles west of Albuquerque. Opened fall of 1998.

Hippocrates Health Institute. 800-842-2125. Tel. 561-471-8876, Fax 561-471-9464. 1443 Palmdale Court, West Palm Beach, FL 33411. http://www.hippocratesinst.com/

All Life Sanctuary. 800-927-2527 ext 00205. Fax 501-760-1492. PO Box 2853, Hot Springs, AR 71914. Wheatgrass pioneer Rev. Viktoras Kulvinskas, MS. http://www.naturalUSA.com/viktor/sanctuary.html/

Tree of Life Center. Gabriel Cousins, MD. PO Box 1080, Patagonia, AZ 85624. 520-394-2533, 520-394-2520. Living Foods Lifestyle Center and private consultations with this medical doctor, author and Essene bishop. Fasting, live foods workshops. Not a resort.

Whitaker Wellness Institute. 800-826-1550, fax 949-955-3005. 4321 Birch St., Newport Beach, CA 92660. Holistic doctors outpatient clinic by Julian Whitaker, MD, editor of *Health and Healing Newsletter*. For subscriptions call 800-539-8219. Not a resort.

Seed Sources

Natural Food Stores. Two major brand names to look for in health stores are *Now Foods* and *Arrowhead Mills*. Both distribute a broad line of grains and seeds that are widely available. The author tests a full line of sprouting seeds for *The Sprout House*. *Sproutman's Sprouting Seeds* are available via mail order and at many natural food stores.

The Sprout House. 1-800-S-P-R-O-U-T-S. See *Mail order shippers.*

Hippocrates Health Institute. 561-471-8876. See *Retreats.*

Optimum Health Institute. 619-464-3346. See *Retreats.*

Good Eats Natural Foods. 800-490-0044. fax 215-443-7087. http://www.Goodeats.com/ Mail order supplier of natural foods including some sprouting seeds. Wholesale and retail.

Johnny's Selected Seeds. RR 1, Box 2580, Foss Hill Rd, Albion, ME 04910. 207-437-4357, fax 800-437-4290. Organic garden and sprouting seeds. http://www.Johnnyseeds.com

Jaffe Bros. Organic foods & seeds. 619-749-1133, fax 619-749-1282

Walton Feed. P.O. Box 307, Montpelier, ID 83254. Emergency food supplies and sprouting seeds. http://www.waltonfeed.com

Juice Bars

Jamba Juice. 1700 Seventeenth Street, San Francisco, CA 94103. 415-865-1134, fax 415-487-1143. A fast growing chain of 130 stores in the Western USA. All stores carry wheatgrass.

Zuka Juice. 477 East Winchester #205, Salt Lake City, UT 84107. 801-265-8423, fax 801-265-3932. http://www.ZukaJuice.com/ This chain has approximately 120 stores in operation in the west, mid-west region. Every store has fresh squeezed citrus and carrot juices and an ounce of wheatgrass juice costs only 94 cents.

General

Rhio's Raw Energy Hotline. 212-343-1152. A raw/live foods help line and resource directory of classes and events in the New York metro area and beyond. Ask for Rhio's new raw foods cookbook.

San Francisco Live Food Enthusiasts. The *Sproutline* 415-751-2806. San Francisco, CA. Telephone help line and listing of live foods pot-lucks, lectures and outings in the San Francisco area.

RawTimes. An excellent website resource for testimonials, e-mail forums, recipes, restaurant reviews, networking, events and book reviews on living foods diet. Http://www.rawtimes.com/

Loretta's Living Foods. Consultations on living foods and wheatgrass therapy by an experienced teacher who worked with Dr. Ann Wigmore. Tel. 610-648-0241

Sproutman. The author gives private consultations on health, healing, diet and fasting. Tel. 413-528- 5200. Fax 413-528-5201. E-mail: Sprout@Sproutman.com Website: http://www.Sproutman.com

AgriGenic Food Corp. 800-788-1084, fax 800-788-1083. 5152 Bolsa Ave #101, Huntington Beach, CA 92649. http://www.agrigenic.com This company manufactures live enzyme supplements from wheat sprouts (pre-wheat grass stage) which are rich in superoxide dismutase.

International Food Allergy Association. Eileen Yoder, Ph.D. 17635 Windsor Parkway, Tinley Park, IL 60477. 708-633-0434.

Other Books about Grass

Cereal Grasses. by Ron Seibold. Published by Keats Publishing PO Box 876, New Canaan, CT 06840. 800-858-7014, fax 203-972-3991.

The Wheatgrass Book. by Ann Wigmore. Published by Avery Publishing Group, Garden City Park, NY 11040. 800-548-5757

The Wheatgrass & Sprouting Journal. A newsletter. 3353 South Main, Suite 197, Salt Lake City, UT 84115. Phone: 801-435-8731, fax 435-783-6401, E-mail: peterg@gamesman.com. Peter Giordano, editor.

Survival Into the 21st Century. by Rev. Viktoras Kulvinskas, MS. PO Box 2853 Hot Springs, AR 71914. Also **Sprout for the Love of Every Body** by Viktoras. Both available through *Twenty First Century Press*, Fairfield, IA 52556. Tel. 515-472-5105, fax 515-472-8443.

Living Foods For Optimum Health. by Brian Clements with Theresa Foy DiGeronimo. Prima Publishing. 916-632-4451.

The Hippocrates Diet. by Ann Wigmore. Published by Avery Publishing Group, Garden City Park, NY 11040. 800-548-5757

How I Conquered Cancer Naturally. by Edie May Hunsberger. Production House, 4307 N. Euclid Ave. San Diego, CA 92115

International Biogenic Society. I.B.S. P.O. Box 849, Nelson, B.C. Canada V1L 6A5. Publisher of the *Essene Gospel of Peace* and other books by Edmund Bordeau Székely.

The Awareness Institute. 1305 South Highland Park Drive, Lake Wales, Florida 33853-7471. 941-676-6231. Fax 941-676-4261. http://www.awarinst.com Mail order distributor of (above) I.B.S. books.

Professional Associations

American Holistic Medical Association. McLean, VA. For patient referral directory, call 703-556-9728, or Fax 703-556-8729. E-mail: HolistMed@aol.com http://www.ahmaholistic.com/

International Sprout Growers Association. Professional trade association for commercial sprout and grass growing. 800-448-8006. Fax 413-253-6965. http://www.ISGA-Sprouts.org

International Association for Colon Hydrotherapy. San Antonio, TX 78230. 210-366-2888, Fax 210-366-2999. E-mail: iact@healthy.net

Kamut Association of North America. http://www.kamut.com P.O. Box 691, Fort Benton, MT 59442. 1-800-644-6450. fax 406-622-5439.

The Invisible Gardener's Natural Pest Control Center. Provides help on Alternatives to Chemicals for the Home and Garden, Farmer and Professional. http://www.invisiblegardener.com

Alliance for Natural Health. P.O. Box 4035, Hammond, IN 46324. 708-974-9373. Fax: 708-974-6002. Formerly Price Pottenger Inst.

Index by Subject

Index by Word

Other Books

by Steve Meyerowitz

Sproutman's
Sprout Chart

Juice Fasting
& Detoxification

Food Combining
& Digestion

Sproutman's
Kitchen Garden Cookbook

Sprouts
The Miracle Food

Who Is This Sproutman?

Steve Meyerowitz began his journey to better health in 1975 to correct a lifelong chronic condition of severe allergies and asthma. After almost 20 years of disappointment with conventional medicine, Steve restored his health through his own program of purification, lifestyle adjustment, exercise, fasting, juicing and living foods.

Over the years, he has lived on and experimented with many so called "extreme" diets including raw foods, fruitarianism, sprouts, vegan–vegetarianism and fasting. In 1977, he was pronounced "Sproutman" by *Vegetarian Times Magazine* in a feature article that explored his innovative sprouting ideas and recipes.

After 10 years as a music and comedy entertainer, he undertook a complete lifestyle change for his health. In 1980, he opened *The Sprout House*, a "no-cooking school" in New York City. There, he began a formal program of teaching kitchen gardening and the preparation of gourmet sprouted and vegetarian foods. Steve has invented two home sprouters, the *Flax Sprout Bag* and *the Sprout House Tabletop Greenhouse* and founded The Sprout House, a company supplying home sprouting kits and a full line of organic sprouting seeds. He has since sold the company to devote himself to writing and educating on the subjects of health and diet.

Steve has been featured on the *Home Shopping* and *TV Food Networks,* and in *Prevention, Organic Gardening* and *Flower & Garden* Magazines. In 3 minutes on QVC, 953 people ordered his Cookbook and Tabletop Greenhouse.

Steve and his family, including two little sprouts, now live and breathe fresher air in the Berkshire mountains of Massachusetts.